Academic Scientists at Work

Second Edition

T0225896

Academic Scientists at Work
Second Edition

Jeremy M. Boss
Department of Microbiology and Immunology
Emory University School of Medicine
Atlanta, GA

Susan H. Eckert
Associate Dean for Administration
Emory University School of Nursing
Atlanta, GA

 Springer

Jeremy M. Boss
Department of Microbiology
 and Immunology
Emory University School of Medicine
Atlanta, GA
boss@microbio.emory.edu

Susan H. Eckert
Associate Dean for Administration
Emory University School of Nursing
Atlanta, GA
seckert@emory.edu

Library of Congress Control Number: 2006920911

Additional material to this book can be downloaded from http://extras.springer.com.

ISBN-10: 0-387-32176-4
ISBN-13: 978-0387-32176-9

Printed on acid-free paper.

9 8 7 6 5 4 3 2 1

springer.com

TABLE OF CONTENTS

INTRODUCTION

Welcome to the world of Academic Science. Academic Scientist careers are challenging, ever changing, exciting, and can be extraordinarily rewarding. The career path requires a long-term education commitment that focuses on asking the right questions, outlining the right experiments, performing experiments to perfection, and presenting the information to the scientific community. The career path also requires that you become proficient in procuring funding for your research projects and maybe your own salary. Unfortunately, most training programs do not focus on how to effectively and efficiently manage a lab or a career. Because scientists entering an academic career find themselves as independent businesspersons, poor career and lab management choices have pronounced effects on their ability to succeed. The beginning Academic Scientist needs a working knowledge of how the system functions and what is expected. This book attempts to provide the reader with a working knowledge of how academic science is conducted, how to approach the various tasks of academic life, and how faculty members are evaluated.

The chapters take the reader along the academic path beginning from the near completion of the postdoctoral fellowship through the promotion and tenure process of the Assistant Professor. There are five sections. Using an analogy to horse racing, *Part I — The Starting Gate* focuses on beginning a career as an Academic Scientist. Approaches to seeking and negotiating a job, managing a lab, writing grants, and interacting with colleagues are presented. *Part II — Down the Stretch* presents approaches to being successful in the three major areas by which Academic Scientists are reviewed: Scholarship, Teaching, and Service. Dubbed the *Finish Line, Part III* focuses on the promotion and tenure process. Included in this section is the chapter *Survey Says*, which presents the data and information collected from a national survey of Academic Scientists about the academic processes discussed in the book. Comments from survey respondents are provided in this chapter, as well as throughout the book with the "Survey Says" heading.

This second edition includes updates to all of the topics presented. Following publication of *Academic Scientists at Work*'s first edition, the authors have been writing career advice columns for *Science*'s *Next Wave* on-line website. Because the topics covered by these articles parallel the mission of this book, these 12 columns are now included in this second edition as *Part IV — The Extras Section* (*Daily Double*) to provide additional insight and depth to the reader. Of course the reader can go directly to the *Next Wave* site to read or print the articles directly (http://nextwave.sciencemag.org/cdc/index.shtml).

A diverse set of appendices comprises *Part V* (*The Winner's Circle*). Included are a series of worksheets designed for the reader to ask and, of course, answer many of the questions that come up when looking for a job and making lab and career management decisions. Sample letters, curriculum vitas, Specific Aims pages, which are diagrammed to explain how they were composed, are also included. Blank copies of the worksheets and databases are included in the CD-ROM that accompanies this book. All of the above documents were formatted as Microsoft Word™ files. Some contain fillable text to allow the user to "tab" through the document and fill in the fields. Microsoft Excel™ files to aid in budgeting are also included. Additionally, six FileMaker™ Pro databases are included which are designed to organize and keep track of a variety of reagents in the laboratory. Each of the databases has been used in the authors' laboratory, some for over 15 years.

THE AUTHORS

Jeremy M. Boss, PhD

Dr. Boss received his bachelor's degree from the State University of New York at Albany. He joined the graduate program in the Biology Department of that school and completed his doctoral dissertation with Dr. Richard S. Zitomer on the regulation of the yeast cytochrome c genes. He continued his training at Harvard University in the Department of Biochemistry and Molecular Biology with Dr. Jack L. Strominger. During this time period, he participated in projects to clone and characterize human major histocompatibility complex (MHC) class II genes. Towards the end of his training with Dr. Strominger he began to study the regulation of MHC class II genes. In 1986, Dr. Boss joined the faculty of the Department of Microbiology and Immunology at Emory University. In 1992, he was promoted to Associate Professor with tenure and in 1997 to Professor.

Scholarship: Dr. Boss' research focuses on understanding the molecular mechanisms that regulate immune system genes. He has contributed to our understanding of how the MHC class II genes are regulated and how genes are regulated by tumor necrosis factor. Dr. Boss has published more than 85 peer-reviewed research articles and has been funded from federal agencies for the last 18 years.

Teaching: Including his current students and fellows, Dr. Boss has supervised the training of 18 graduate students and 16 postdoctoral fellows. He has served on 71 PhD thesis committees (YIKES!). Dr. Boss has taught in a variety of immunology and genetics related graduate and medical school courses, including one that he created called EGOR (Eukaryotic Gene Organization and Regulation). He was also the course director for the

Microbiology and Immunology course provided for the physician assistant program at Emory for seven years. He is currently the course director for an introductory immunology graduate course.

Service: As an Assistant Professor Dr. Boss became involved in the administration of the bioscience graduate programs at Emory. This experience began as a graduate student recruiter for his departmental program. He was then elected by his colleagues to serve as the Director of Graduate Studies for an interdepartmental program in Genetics and Molecular Biology. After four years at that post he was elected to serve as the Program Director of the graduate program in Genetics and Molecular Biology and has served in this post for 9 years now. Dr. Boss also served on over 50 committees in the medical school, including membership on the ad hoc Tenure and Promotions committee and later as a charter member of the standing committee on Tenure and Promotions. In the School of Medicine, he has also served on the Research Advisory Committee to Dean's office, as well as on the Postdoctoral Fellow Advisory Committee.

Dr. Boss has also served on a variety of grant review panels. These have included membership status on two American Cancer Society review panels and an NIH review panel. He has also served as an ad hoc reviewer for the NIH over the last 15 years on various regular and special study section panels. He reviews approximately 20 manuscripts each year for a number of journals. Handling more than 15 manuscripts a week, Dr. Boss is currently a Deputy Editor for his field's society journal: *The Journal of Immunology*.

Susan H. Eckert, PhD

Dr. Eckert is Associate Dean for Administration at the Emory University School of Nursing, and has responsibility for the School's finance, grants administration, research compliance, human resources, and administrative planning activities. She participates in a wide range of University services, including faculty and administrative search committees, sponsored program compliance groups, human resource policy committees, financial review groups, and other committees that exist either to relieve or beleaguer Academic Scientists. Dr. Eckert is a member of the Board of Directors of DeKalb Technical College in Clarkston, Georgia, and is on the Board of Directors of the Emory University Credit Union.

She received her doctorate in Higher Education Policy from Georgia State University, where the focus of her doctoral research was leadership factors affecting research-intensive basic science departments in medical schools. She learned everything she knows that is worthwhile about the business of science and the values of a meritocracy during two decades as administrator of the Department of Microbiology and Immunology at Emory University.

CAST OF CHARACTERS

Throughout, the authors have used (or attempted to use) humor to portray events, situations, and problems that occur in the life of an Academic Scientist. While the sections may appear funny (or not), they all have a point and all address important issues.

Additionally, a number of fictitious characters were created to play specific protagonist and antagonist roles in this book. While the authors' colleagues have tried to guess whom they represent, we are sorry to say that they do not represent anyone in particular. The characters are the best and worst of all scientists' personalities and behaviors. The characters include:

Dr. Ima Starr — is the rising superstar looking for her first faculty position. She will encounter all of the problems associated with managing her career and laboratory discussed in the book. As described in her CV and Specific Aims, she studies the biology of Ohmygaud and Blahdeblahs.

Dr. Kneematoad — is the major antagonistic character. He exists in all walks of life. Dr. Kneematoad's ambitions are to make everyone do his work while promoting his own goals and abilities at every opportunity. In case you were wondering, he does not work on nematodes.

Dr. Mary Musculus — is the caring chair of Dr. Starr's department. Her role is to make sure that the faculty are supported and mentored so that Dr. Starr and all her faculty can be the best that they can be.

Drs. Brillodooz, Pat I. Ence, Rekinwith, and Dewit Miweigh — represent everyday colleagues with the usual science personalities and hairstyles.

The graduate student with the curly red hair — is Dr. Boss as his own worst nightmare.

There are other students, fellows, and research technicians whose names appear in the appendices strictly for amusement.

LEXICON

This glossary provides definitions to some of the terms (real and made up) used in the book. The terms are arranged by subject. The chapters where the term was used most are indicated.

Tenure-Related Terms:

Clinical Track (*Chapters 1 & 10*): Faculty level appointment for physicians or physician scientists. Clinical-track positions are typically renewable each year and do not usually have tenure associated with them.

Research Track (*Chapter 1*): Faculty level appointment for researchers that is renewable each year. Research track appointments are not eligible for tenure. Research-track positions are used throughout academia. Often such positions allow young investigators the opportunity to apply for and procure their own funding.

Tenure (*Chapters 1, 2, & 10*): Thanks to the continuing efforts of the American Association of University Professors (AAUP), tenured faculty in higher education have a degree of certainty regarding their continued employment. The award of tenure is a significant event because it usually means that the faculty member has a job until retirement or resignation from that college or university. Tenure is an important part of the system of values in colleges and universities not only because of the sense of security that tenured faculty enjoy, but also because it signifies that the faculty member has been evaluated by his or her peers and found to be a valuable colleague. Tenure can be lost if the faculty member misbehaves egregiously, resigns and goes to another institution, or if the University goes broke.

Tenure Track (*Chapters 1 & 10*): This is the "path" to a tenured position. If a position is described as tenure track, then the incumbent needs to know how long she can remain "on track" and stay employed before getting tenure. It is important to know whether or not a position can lead to tenure because only positions that clearly state that they are tenure track can be assumed to lead to tenure. The time to tenure varies with each institution.

Tenure Clock (*Chapters 1, 2, & 10*): Positions that are on tenure track also are "on the clock," because there is a finite time period for a junior faculty member to establish his or her value to the institution. Each university has its own policies about when the clock starts ticking after the initial appointment, and when the alarm sounds that the time has run out.

Up or Out (*Chapters 1 & 10*): This term relates to the tenure clock and when the alarm will go off that the time on the tenure track has run out. The policies of the university are usually very clear about this. For example, an Assistant Professor on seven-year tenure track may be able to stay in that position for a total of 7 years. Because faculty usually get an entire year of notice that their jobs won't be continued, and because it takes nearly a year to conduct the tenure evaluation, it means that the tenure review must be completed for the Assistant Professor no later than the end of the 6th year in tenure track. That is the move "up" or get ready to go "out" year.

Job-Related Terms:

Job Talk (*Chapter 1*): The formal seminar presentation given by a faculty candidate on his or her first visit.

Chalk Talk (*Chapter 1*): Job candidates may also be asked to present a less formal talk to the faculty on either their first or second visits during the interview process. While the setting of the chalk talk may be more casual, the content of the talk typically focuses on current work in progress and future goals and plans.

CV (*Chapter 1*): Curriculum Vitae. This is the paper representation of an individual's career in academe. In the non-academic world, this document is called a resume, but is much less detailed. The CV presents the specifics of academic work, including education, publications, funding, and presentations, a chronology of academic jobs, research interests, service activities, memberships in professional organizations, etc. Keeping this document up to date helps an Academic Scientist remain prepared for opportunities and saves time.

The Macon Test (*Chapter 1*): Originally coined by our colleague Dr. Gordon Churchward, the Macon Test is the mental calculation a faculty member employs to determine if he or she would like to have a faculty candidate as a companion on a three-hour road trip to some town like Macon, Georgia (scientists in the Northeast might use Poughkeepsie as the destination). This quick test is based on our first impressions of the people we meet. In this case, we, the faculty, ask ourselves if the prospective candidate would be good company over the long haul of an academic lifetime, or if he would be an annoying or boring passenger on the trip. In the latter case, the candidate flunks the Macon Test, and might not get a job offer.

Parking Committee (*Chapter 9*): Although serving on such a committee is a thankless job because someone's gotta do it, junior faculty members should avoid at all costs serving on committees that do not focus on research.

Second Visit (*Chapter 1*): This term refers to the pleasant occurrence when a faculty candidate receives an invitation to return to a campus for another series of interviews (or perhaps even to talk to a real estate agent!). Being invited for a second visit signifies that the candidate's name has probably been put on the "short list" among the other candidates.

Grant- and Money-Related Terms:

Banking and Discretionary Funds (*Chapters 1 & 4*): Many universities provide funded investigators some return on the indirect costs that are generated by their research grants. Regardless of how the funds are generated, the money typically can be carried forward to future years and may be unrestricted in its use to support the faculty member's research program.

Carry Forward (*Chapter 4*): Funds that can be transferred from one fiscal year to the next.

Direct and Indirect Costs (*Chapters 1 & 4*): These terms apply to extramural grant money. The "direct costs" are for the investigator and the direct expenses of the project, and the "indirects" are for the institution. At some universities there may be a plan for funded investigators to share in the indirect costs (see banking and discretionary) and use a portion of these funds to support the specific project or other projects.

Funding/Granting Agencies (*Chapter 3*): Public and private organizations that accept applications for grants. Each has its own regulations, application forms, and calendars of deadlines and award dates. Federal granting agencies include the National Institutes of Health (NIH), National Science Foundation (NSF), Department of Defense (DOD), etc. There are scores of private foundations and organizations that have grant funding programs, including the American Cancer Society; American Heart Association; American Lung Association; Multiple Sclerosis Society; AMFAR (American Foundation for AIDS Research), etc.

Grantsmanship (*Chapter 3*): The skills associated with the grant application and award processes. These skills are learned, sometimes through trial and error experiences, but can be taught to those Academic Scientists who are open to the advice of their experienced colleagues.

Hard Money vs. Soft Money (*Chapters 1 & 3*): Hard money refers to the annually recurring funds that come from the University, via its endowment income or, in state institutions, from the state government. Soft money refers to the funds that are garnered from grants or contracts, and which have to be competitively renewed to continue past their initial term. Normally, Academic Scientist salaries are composed of a combination of both hard and soft money to reflect the dual commitments both to the university (teaching and service) and to the research enterprise.

HIC, Human Investigations Committee and Institutional Review Boards (IRB) (*Chapter 3*): These are standing committees of the university or institute that are responsible for reviewing all protocols and procedures that involve human subjects in research. No research involving humans at a university or other institution can be undertaken until these committees approve the procedures. The sanctions for proceeding without such approval can include being banished from receiving future grant funds.

IACUC, Institutional Animal Care and Usage Committee: This is also a standing committee in a research institution. It reviews all the protocols and procedures that involve the use of animals in research. The committee consists of scientists, veterinarians, and lay people.

Principal Investigator (*Chapter 3*): The lead investigator of a grant. A grant can only have one Principal Investigator, but may have several co-investigators.

Program Director (*Chapter 3*): This term has several meanings. A program director is the principal investigator of a multi-component grant, such as a program project or training grant. Program Directors are also the officials at the NIH who coordinate the extramural funding programs. They are the people who will award you money.

Request for Applications (*Chapter 3*): Funding agencies announce "Requests for Applications (RFAs)" with specific information about a topic/disease/system that the agency is targeting for funding. Submitting grant applications in a targeted area of funding is a very good way to obtain money from federal sources in such fields as infectious diseases and bioterrorism. Information about upcoming RFAs is available through the websites of the funding agencies and from the university-sponsored programs offices.

Streamlining (*Chapter 3*): This is the term used by the NIH for the process by which 50% of the grant applications are removed from the number of grants to be discussed by a review panel/study section. This process, which used to be called "triaging," allows the group to focus on the most competitive grant applications, and returns the least competitive applications back for retooling.

Study Section or Peer Review Group (*Chapter 3 & 6*): A meeting where grant reviewers convene to discuss and rank a series of grant applications.

Research and Publication Terms:

Fishing Trip (*Chapter 8*): This is the informal term for high-risk projects in which the investigator is searching for a small number of items among a very, very large number. If successful, fishing trips can lead to significant discoveries that have a big payoff. If the fishing trip yields no "catches," then the cost of materials and the people-power that was lost to the trip are the downsides. Also, if a graduate student is sent on such a trip, with the idea that it could result in a spectacular thesis project, then it should be understood that an unproductive fishing trip could delay the student's completing his degree. Therefore, knowing when to "punt" is a valuable trait.

Magnum Opus Syndrome (*Chapter 6*): When an investigator delays publishing his or her work until an entire system is solved. While complete stories are favored, waiting too long or waiting until your work is good enough for the Top-of-the-Heap journals can be a problem for the young investigator on a tenure clock. A side effect of this syndrome is Bin Scooped Disease.

Minimal Publishable Unit (MPU) (*Chapter 6*): An MPU is a manuscript that contains the smallest number of experiments that tests the hypothesis or answers the questions that are posed. While one should not fill one's CV or scientific reputation with MPUs, sometimes the data and results are clearly an MPU and should be published as such. A compromise between the Magnum Opus Syndrome and an MPU is recommended.

Ortholologous Protocols (*Chapter 4*): This is the term that describes the situation when a lab uses a wide variety of protocols to accomplish the same goal. This can be problematic, because having too many protocols for the same purpose usually leads to many variations in the quality of the data or reagents being produced. It is better to have a small number of protocols that everyone uses so that they can be more readily evaluated when experiments do not work.

Pink Sheet (*Chapter 3*): Now called the Summary Statement, the pink sheet is the documentation of the review of an NIH grant application. The pink sheet (which used to arrive by mail on pink paper) includes the score of the application, the percentile ranking, and a narrative synopsis of the comments made about the grant application by the study section reviewers. Nowadays, the summary statement arrives as a PDF in your email. But you can still print it on pink paper if you so desire.

Protocol Drift (*Chapter 4*): A procedure that becomes altered as it is passed down through the personnel of a laboratory. This becomes an issue when the "drift" takes a perfectly good procedure through an evolutionary process that makes it a procedure that doesn't work at all. Having written protocols for laboratory assays and procedures and implementing a policy that prevents lab members from casually changing the protocols is the remedy to protocol drift.

Punting (*Chapter 6*): A strategy used both in football and in research. In research terms, punting occurs when a project is deep in the hole and the Principal Investigator realizes that it is time to abandon the project and move on to other more promising projects. The key issues in deciding whether or not to punt are: Why? When? and How?

Society-Level Journals (*Chapter 6*): These are peer-review journals that are published by the various scientific societies and represent that specialty of science. Work published in these journals is considered important for the advancement of a field.

Sub-Specialty Journals (*Chapter 6*): These are peer-review journals that are field oriented. They represent the largest class of journals. Work published in these journals will have a lesser impact on the broad discipline that they represent.

Top-of-the-Heap Journals (*Chapter 6*): This refers to the small number of journals that have the highest academic ranking and prestige. Publications in these journals are cited frequently.

'Tweener (*Chapters 6 & 10*): All middle authors on scientific papers. Irrespective of what they actually did to justify their names on a scientific paper, 'tweeners do not get full credit for the work.

Lab Management and Other:

Bench Warp Shuffle (*Chapter 4*): The movement of individual workspaces within the laboratory to increase productivity. This drastic maneuver can separate two contrasting personalities, or remove one individual from an environment where he/she is overly distracted or distracting.

DuPlex (*Chapters 1, 2, and 4*): A large scientific instrument company or a two-family house.

GammaZoid (*Chapters 1, 2, and 5*): Name given to an essential laboratory instrument with no known function. A GammaZoid is the latest model of the former BetaZoid.

House of Pain (*Chapter 8*): From the student's point of view, the house of pain is the practice of actually sitting there and watching his mentor edit his paper. While the process is slow and potentially boring, it is an excellent way for trainees to learn how to write and edit a manuscript. This phrase was coined by students in Jack Strominger's group some 20 years ago.

Solo and Mini Groups (*Chapter 4*): are approaches to organizing and assigning research projects to the members of your laboratory. While the solo strategy assigns separate projects to each lab member, the mini-group approach attempts to take full advantage of the strengths of the small laboratory by organizing the tasks of a project through the use of groups. Most labs use a combination of both strategies.

TriPlex (*Chapters 1, 2, & 5*): A smaller scientific instrument company or a three-family house.

ACKNOWLEDGMENTS

Many of the problems and issues that are addressed in the book have been encountered by the authors over the years. We thank the individuals for creating these issues, as well as ALL of the faculty and administrators that have provided advice on how to solve them. In particular, we thank Drs. Charles Moran, Gordon Churchward, Dean Danner, Steven Warren, and John Spitznagel for their eagerness to provide insightful comments on life in science. We also thank the many individuals who have provided comments on the various chapters of the book. Special thanks go to Dr. James Austin, who served as *Science*'s editor for our *Next Wave* articles. Lastly, we are grateful to those scientists who took the time to answer our survey or to participate in one-on-one interviews with us.

Good luck and enjoy!

PART I

THE STARTING GATE

Starting with the process of looking for and negotiating an academic position, the chapters in Part I focus on beginning a career as an Academic Scientist. Issues discussed include setting up a lab, writing and submitting grant applications, managing a laboratory, and contributing to the academics of a department. A number of worksheets and database files are included in the Appendix and CD for use with the chapters.

CHAPTER 1

GETTIN' A JOB

Welcome to the world of biomedical research. Over the past quarter century, biomedical research opportunities have skyrocketed. Literally hundreds of companies have opened their gates for biomedically trained scientists and universities throughout the world have expanded their research programs. They are all competing for you to work for them. You need to decide which type of job you want and what career path to take. After deciding on the type of job you want, you will need to find a job, get an offer, and negotiate the terms of your employment. This chapter will focus on these issues and includes:

- Jobs: types and when and how to apply
- Preparing for your interview
- The second visit
- Negotiating your position

JOBS

Types of Biomedical Research Positions

There are numerous types of positions. A major choice is to decide between academia and pharmaceutical or biotechnology companies. To decide what type of position to pursue, you should begin by asking yourself and honestly answering questions relating to your own abilities; your likes and dislikes; and how you want to spend your working life. Here are some examples of questions that will reveal your interest in a career as an Academic Scientist.

- Do you want complete control over your research programs?
- Can you tolerate writing grant applications for research funds?
- Are your writing skills up to par?
- Can you meet deadlines?
- Do you respond well to critical review of your work?
- Can you adhere to a budget?
- Do you enjoy teaching and lecturing?
- Do you enjoy training others in your field?
- Can you tolerate administrative paperwork?
- Do you require job security?
- Do you want to earn a lot of money?

If you answered **Yes** to all but the last question, then an academic career may be for you. If you answered **No** to several or all of the questions, then a job in the biotech or pharmaceutical industry may be a better choice.

Academia vs. BioTech

The biotech industry has changed over the years. The work is challenging and rewarding both on an intellectual level as well as on a financial level. Some corporations have "pure research" divisions that allow scientists the freedom to pursue a general area of research in the hopes that the area will lead to a product down the road. This type of position offers freedom from writing grant applications, teaching, and a host of other responsibilities required in academic positions.

A position in a biotechnology company could also potentially provide you with high levels of funds and resources not available to the typical Academic Scientist. In some situations, the goals of the position are made absolutely clear, and the research direction is laid out over the long term. If you don't mind relatively less control over your research life, are indifferent to job security, dislike writing and submitting grant applications, are indifferent to or dislike teaching, and prefer not to have your work evaluated constantly by anonymous groups, then a biotech position may be perfect for you. In addition, obtaining stock options in an up-and-coming company may be a real bonus to the work.

Academic positions offer different incentives and rewards. Perhaps the greatest benefit of academic life is the freedom to work in any field you choose. The caveat is that you will have to find your own funding for your work, convince others that the work is worthwhile for you to stay in the department (tenure issues), and be willing to receive lots of objective — and maybe even subjective — critiques of your ideas. Freedom has its price.

A second feature of academic positions is independence. While this sounds a lot like freedom of choice, it is not. Each "Principal Investigator" (PI), as the National Institutes of Health (NIH) and other granting agencies call their project coordinators, runs his or her own small business. Thus, independence means that you will be your own Boss. Being a Boss means that within the policies and procedures of your institution you get to run your own show and no one will tell you how to run it differently. However, being the Boss also means that you are responsible for hiring and firing, supervising, directing, and mentoring your staff. It means filing the correct paperwork for licenses and approval of your protocols. Unlike the BioTech firm where you are part of a business, you now ARE the business.

A third feature of most academic positions involves a host of responsibilities that include various types of teaching, ideological discussion and exchange with colleagues, shaping of an educational institution, and development of departmental research goals and initiatives. Teaching and training of students is a major part of academic life. You will be judged to some extent on your ability to teach and the amount that you participate in the education mission of your department. Because academic positions have a wide range of teaching requirements, you should be able to find one that provides you with the amount of teaching responsibility that you desire.

The other items mentioned above fall into the general category of service to your department or division. How much of a team player are you and do you like to take a leadership role? If you want to get involved in running the academic programs at your institution, no one will stop you. In general, faculty get to have a say in the way their institutions are run. At least they think they do and that may be all that matters in the end anyway.

Job security is another benefit for the academic scientist. Most academic research institutions have either tenure or long-term contract policies. The prospect of attaining "a job for life" seems like a pretty good deal, but the process of attainment may not be one you want to go through. Additionally, having a job but not enjoying it would make the "job for life" like a prison sentence.

Thus, freedom, independence, security, and contributing to an academic environment highlight the advantages of a job in academia. While many consider tenured professorships the best of all jobs, the process of becoming a tenured professor is rigorous but can be accomplished with hard work, careful planning, some good ideas, and a little guidance.

Types of Academic Positions

To be inclusive in hiring and to allow individuals to progress through the ranks, academic institutions offer a plethora of "tracks" and levels. Each track (such as *tenure track* or *research track*) comes with a set of promotion levels and its own evaluation system for advancement. Being on the right track when you take your job is important because it may be difficult to switch tracks later. Each academic institution will have its own specific set of rules. These rules are not top secret, although you will find that most junior faculty don't know what the rules are or where to find them (as we found out in our national survey of faculty; see *Chapter 11*). In all cases, the rules are published in some sort of guide or faculty handbook that you may receive when you take the job. Most schools post the rules somewhere in the labyrinth of their webpages, so you may want to take a look at the websites of the institutions you are applying to for positions.

The most important distinction between the tracks is whether they lead to tenure or not. A tenure-track position typically offers the highest degree of independence and allows faculty members to pursue their own research interests. A tenure-track appointment will also require the highest level of accomplishment from a faculty member when the time for the tenure decision arrives. As discussed in later chapters, tenure-track faculty members will be judged on their level of excellence in scholarship, teaching, and service. Tenure-track positions also come with a built-in timeline. The phrases "publish or perish," "up or out," and "sorry Charlie" refer to the necessity to become established within a specific time frame, that is, the time frame of the "Tenure Clock."

Non-tenure research-track positions are also available in many institutions. The independence and evaluation/promotion criteria for this type of position vary. In the best cases, non-tenure, research-track positions allow faculty to apply for funding, run their own laboratories, and train students. In this situation, the possibility of continuous appointment does not exist and employment is likely to be dependent purely on scientific productivity, which may be measured in an individual's ability to generate funds for his entire salary. In the "least best" scenario, a non-tenure research-track Assistant Professor or Instructor position allows you to pursue research funding, but does not come with academic independence. For example, a research-track Assistant Professor's space may be part of a senior, tenured Professor's space. In addition, research-track faculty may not have a say in the future direction of the department. This latter situation often occurs as postdocs become very senior in their positions. This type of position becomes an ideal place for young investigators to begin an independent career because it allows them to test their abilities in carving out their own niche within a larger research program. If successful in procuring funds, the young investigator is now in an ideal position to seek a tenure-track position at another institution.

Clinical-track positions exist in medical schools and other professional schools that have substantial clinical service responsibilities. These tracks are designed for faculty who wish only to perform clinical service and to train students and fellows in the art of medicine. Evaluation of the performance of clinical positions is based mostly on the clinical service that is provided by the faculty member, rather than on the research or scholarly activities. This doesn't mean that scholarly accomplishments (clinical reports, research manuscripts, etc.) may not be required for promotion. Salary funding for clinical-track faculty who need time for research is complex because their source of salary support usually relates to their clinical service, not research. Before a physician scientist takes a job, the service and research time should be clearly defined (see below). This would apply to a tenure-track physician/clinical scientist position as well.

When Should You Look for a Faculty Position?

When you are ready! How do you know you are ready? You are ready when your accomplishments are sufficient for other faculty to believe that you now possess the background, training, and expertise that they wish to have in a colleague in their department. When departments carry out job searches they often are looking for a faculty member in a specific area, such as autoimmune diseases. Other times a department may be looking for the best person they can find in a broad area or field such as immunology. A job ad that is posted in an international journal such as *Science* can generate more than 100 applicants. Some of the applicants will have no relationship to the field advertised, but most will. A departmental search committee will review the applications and regardless of the parameters of the job itself, the committee will weigh the productivity, current funding, training history, and future goals of each candidate. You are therefore "ready" for a faculty position when you have established yourself in an area in which you can successfully develop your own independent research program. You also must have developed ideas of how you will pursue your research program. Because it can take more than six months to get a job after you have applied, once you have decided that you are marketable, you should begin to look at job postings and submit applications.

Departmental Faculty Searches

To understand how to get a faculty position in academia, it may be helpful to view the process from the inside. Departmental search committees are appointed by the departmental chair with the specific task of finding candidates for a particular position. One of the faculty members will serve as the "chair" of the group and will most likely be the contact person noted in the ad. The committee will write the ad, read the applications, and prepare a short list of candidates that they think should be invited. Thus, it is this committee that you must impress first with the strength of your **curriculum vitae** (**CV**). A limited number of candidates are then invited, usually over a short period of time, so that comparisons can be made among them. Ultimately, the hiring decision will be made by the departmental chair. However, her decision will usually be made with substantial input from both the search committee and the rest of the departmental faculty. Each faculty member in the department will have an opinion on who is the best candidate. Their opinions may be influenced by the candidate's field of expertise, productivity, and will no doubt be influenced by the candidate's bubbling personality. Each faculty member will ask the question, "How will hiring this person improve the department?" They may even ask the question, "How will this person improve my own research program?"

How Do You Find Out About a Faculty Position?

Look in the back of journals like *Science*, *Nature*, and *Cell* for job postings. Some journals and certain scientific associations have on-line job websites that you can browse. A list is presented below:

```
www.science.com
www.asm.com
www.the-scientist.com
www.FASEB.org
www.nature.com
www.cell.com
www.scijobs.org
www.sciencejobs.com
```

You should also be discussing this aspect of your career with your mentor, who may just happen to come across someone at a national meeting saying that they are looking for someone like YOU! This happens often, so it's important to be ready to apply when the opportunity arises. You may be approached to apply for a job after you give a talk at a scientific meeting. This also happens with some frequency and therefore you should make every attempt to go to meetings and present your work.

Applying for the Job

The key to getting an academic position is dependent on what you have already done and a prediction of what you will do. Therefore, what you send the search committee must reflect your talents in the most positive manner. Typically the search committee will request your CV, list of personal references, a short research description, and future goals.

Your CV is a record of your professional life. It should include your education, honors, professional appointments, publications, research support, students trained, teaching experience, and scientific service (national and local). Your CV is a dynamic record of your career and, once started, will track your accomplishments and document your research activities. If you do not have one ready, now is the time to begin. The CV of Dr. Ima Starr is included as an example in **Appendix 1-1**.

Whom should you ask to write your letters of reference? Choose scientists who know you well. Labmates and pubmates are unlikely to elicit the same regard on the part of a search committee as will bona fide faculty members. This means that faculty other than your mentor should know who you are and what you do. Collaborating with other labs, presenting your work at departmental seminars, asking questions in seminars, and interacting with others in your department are among the best ways for you to become "known." The more famous the scientist writing the letter, the more weight (hopefully positive) it carries with the committee.

Your cover letter to the department search committee should be very *clear* and *brief*. In addition to acknowledging the ad or the way in which you learned about the job, it should describe where you are currently working, your field of expertise, and what you expect to accomplish in your research program. All of this should not exceed one page. An example of a cover letter is included as **Appendix 1-2**.

Your research summary and future goals should also be clear. A ten-page description will not be read carefully. While brevity is important, you do not want to leave anything out that may be critical to your getting an interview. Describe your postdoctoral research project aims and findings in the first section. In the Future Goals section, outline the major problem(s) that you wish to attack, its relevance to science and disease, and the specific goals that you would accomplish in a 3- to 5-year period of time. This is essentially an expanded Specific Aims page as required for grant applications (see *Chapter 3*). It is important to remember that you are the expert and not everyone on the search committee has your expertise, so do not dwell on specifics. If you have created a novel reagent or have developed a new technique that you will exploit in your future work, be sure to highlight this in your description. Be certain to convey your enthusiasm for your research program and goals.

> • Have your mentor read your research proposal and get his or her comments before sending out the package. This will also provide your preceptor with a small hint that you will not be around forever.

THE INTERVIEW

Congratulations! You just received a phone call from Dr. M. Musculus inviting you to visit her department. You are now an official faculty candidate. You will probably have about three weeks to get ready for the interview. You will need to prepare a presentation, get ready for one-on-one meetings with the faculty, and determine your start-up requirements.

Seminar / Presentation

The first objective is to make your seminar presentation competitive for a faculty position. Nowadays most departments are outfitted with sophisticated seminar computer projection systems. Your only real concern is that the equipment in the seminar room is compatible with your files. If you will need to use their computer be sure that it has enough memory to run your images. Even if you bring your own laptop, you should have your presentation burned onto a CD in case something goes wrong and you need to use their equipment. Just be prepared for the worst-case scenario.

Now that the presentation platform is set, what should you present? The talk should be *no more than* 50 minutes from start to stop in your practice sessions. You will probably start late and you may get interrupted along the way. Not being able to present your "hot science" at the end of your seminar because you ran out of time would be unfortunate. To have a chance at landing the position your talk needs to be clear and organized, not to mention exciting. The perfect talk would have the following:

- **Thanks** to the committee for inviting you
- **A little joke**, which should not offend anyone or anything (dead or alive)
- **Objectives** — what you will tell them
- **Background** that is required for everyone to understand the talk
- **Significance** of the problem and what needs to be done
- **Experimental** section #1: goal/hypothesis/question, approach, results, and interpretation/conclusions
- **Experimental** section #2: goal/hypothesis/question, approach, results, and interpretation/conclusions
- **Etc ...**
- **Model** of system derived from your work and others
- **Conclusions**
- **Future goals and aims,** with significance and relevance to life science
- **Acknowledgements**

All images that you use in your seminar should have a light-colored or white background so that the overhead lights in the room can be on. While you may get your job if your future chair falls asleep, it is not the best strategy. If you insist on dark backgrounds like blue on your visual media, **DO NOT** use red lettering. Sorry to dwell on this, but no one can see red on blue clearly even if they are not colorblind. Remember: LIGHT AND BRIGHT. Looking at the color scheme of your slides beforehand with your labmates is the best approach. **Appendix 1-3** on the ASAW CD provides examples of good and bad color combinations and some other tips on PowerPoint presentations. Be sure to spell check everything. Misteaks look sloppy, dont' they?

Our *Next Wave* article **The Job Talk** (reproduced in *The Extras Section*) presents a more lengthy discussion of this subject and a series of Dos and Don'ts to be mindful of when you are making your presentation.

You may find that you are rolling along in your actual job seminar, and all of a sudden Dr. Kneematoad interrupts from the audience and begins to ask a bunch of questions. Before you know it, there are 27 other questions, most of which are off the point of your seminar. This is the time for you to take control of the talk. Answer the first few questions and state that you would be happy to discuss these important details after the talk, but to fit into the time frame you need to move on. This will often work and allow you to continue. Because no one likes to be in a seminar for more than one hour, keep track of the time. Many people put their watches on the podium so that checking the time is not obvious.

> ▪ Aside from the quality of your science, the next most important aspect about the presentation is that you should enjoy giving it and show it. Your enthusiasm will be contagious. Good luck and go get 'em!

Meeting with the Faculty

There are the two perspectives in meeting with faculty at the institution when you are interviewing: Theirs and Yours. You will meet with faculty in both formal and informal settings during your visit. These sessions will occur during meals, in their offices, and in group sessions.

From Their Perspective

The faculty, chair, and search committee will be trying to determine what your research plans are and how well you can communicate your ideas. Therefore, you need to have a general research plan and goals. Because you've addressed this in your initial job application, you should be able to use your research statement from your application as a guide and discuss this with the faculty. You may be asked about what you think you could teach. Before you say that you could teach gross anatomy, be sure that is what you want to do. You may get stuck with more than you want.

It is in these interactions that the faculty will decide whether you pass the "Macon Test (see *Lexicon*)." The Macon Test is a top-secret test that each faculty member uses to evaluate whether or not you would be a compatible colleague. So, if you are bored, yawn seven times, and your eyes glaze over when they are telling you about their research, you failed. If you do not listen to what they are saying and insist on doing all the talking, you may also fail. The key here is to be interested in your potential colleagues, exchange ideas with your colleagues, and find out about this place that they call home. If you truly aren't interested in what they have to say, then this may not be the job for you.

From Your Perspective

You have two goals. The first is to make sure that you get an offer. To improve your chances of receiving an offer, your science seminar must be well rehearsed and smooth.

You also need to be able to interact easily with the faculty. To achieve the latter, look at the websites and publication records of the faculty before your visit. You may even consider reading some of their papers. This will allow you to talk with them about their research. Because most candidates don't bother to do this, the faculty will be impressed that you know who they are, what they work on, and that you took the time to do your homework.

The second goal is to find out what the job will really entail. In these sessions, you need to find out about the department and its organization, the graduate programs, clinical responsibilities of the department, the institution, the students and fellows, and about the city. Therefore, these interactions are critical for you to obtain all the information that you need to know before accepting the position. The worksheets in **Appendices 1-4**, **1-5**, and **1-6** are designed to help you sort through the many academic issues that you should consider. The worksheets are broken down into the three general areas: **scholarship**, **teaching**, and **service**. The information that you gather should focus on answering a simple question:

> Where will my career be in 5 or 6 years (tenure evaluation time) if I join this department?

You should ask the same sets of questions to several people, as their answers may be different. It will also give you something to talk about in case you encounter a lull in the conversation. When meeting with junior faculty, try to get a handle on their overall satisfaction and whether or not all of their expectations have been met.

The Research Environment Worksheet (**Appendix 1-4**) focuses on the current and future research atmosphere of the department and institution and how easy it will be to be able to carry out your experimental plan. The key issues include:

- Is the department growing?
- How many other faculty are being hired by the department or institute?
- What is the distribution of senior and junior faculty?
- How many people would you be able to collaborate with and learn new technology from?
- Are there shared resources/equipment?
- What are the core facilities?
- How many postdocs are in the department?
- How many graduate students are in the department?
- Is there a training grant associated with the department?
- How are student stipends paid?

The Teaching Worksheet (**Appendix 1-5**) focuses on the general teaching enterprise that the faculty in the department provide. Here, the key issues are the types of courses, number, and frequency. The types of courses are: undergraduate courses, graduate courses, and professional courses — those designed for medical students, physician assistants, or nurses. As you will read in *Chapter 7: Being a Teacher*, the time commitment involved in preparing for each of these course types is vastly different. The relative teaching load of the other junior faculty will provide you with a gauge as to how much "real teaching" you will be asked to do once you are hired.

The Service Worksheet (**Appendix 1-6**) centers on the other aspects of your position. In addition to research and teaching, the position may have a specific service requirement. If you are an MD, and the department that you are looking at is a clinical department, your job may include seeing patients. Alternatively, you may have to run a clinical lab or core facility. Remember, providing a service requires your time. This issue is critical to physician scientists who wish to establish a research career. If the department is a basic science or an undergraduate-focused department you may have to serve as an advisor, mentor, or have some other administrative responsibility. Because there are only 168 hours in a week (if you worked around the clock), you will need to find out how others in the department are allocating their service time.

While it is strongly encouraged that you use the worksheets, we suggest that you use them to prepare for your visit and fill them out in your hotel room or at home. This will allow you to focus on averaging the data that you receive from the faculty and jotting down the consensus view rather than that of one faculty member or another. Additionally, if you can't remember the answer to a series of questions back in your hotel room, it is likely that you have to ask the question again.

> ▪ The worksheets should be used as a guide to obtain information and help you evaluate the different jobs that you will ultimately compare. You will find that all jobs are different and the one that agrees the most with your ambitions and personality will be the one that will allow you to achieve the most out of your career.

THE SECOND VISIT

Congratulations! You are being asked to come for a second visit. What is the purpose of a second visit? A second visit means that they are very serious about having you join their faculty. As in the first visit, there are several goals depending on which side of the fence you're on. From the faculty side, they want to make sure that you are The One from a scientific point of view, as well as from the Macon Test point of view. They also want to impress you with the quality of the local research and resources, as well as with the living environment of the city. They will be in "recruiting mode." From your point of view, this is your opportunity to: ask all the tough questions, begin to negotiate your position, and decide if this is the area of the world in which you wish to live. Your visit is likely to be broken down into: (a) meeting with faculty in the department, faculty in other departments, or even the Dean; (b) giving an informal presentation; and (c) being shown around the area. Your spouse or significant other may be invited to join you on this trip.

You will likely be scheduled to meet with those who are most anxious to have you join their department and you will meet faculty who missed your first visit. You may want to know if there are faculty outside of the department whose research interests you. Because the potential to interact with these people may be critical to your success, finding out the breadth of scientific expertise at the institution is important. Try to set this up ahead of time so that you can meet with these individuals. These are good opportunities to obtain more details on the questions you asked during the first visit. If you have been to other places to interview since that visit, you will now be able to "comparison shop." Make the most of your time.

You may be asked to give an informal presentation about your research that is aimed at what you will do in the future. Because these "chalk talks" are less structured there is a tendency to under-prepare. What do they want to hear? First, the faculty want to be excited about your research. If there were some naysayers on your candidacy, they need to be convinced that everyone else was right in being interested in you. So, the talk must be well organized. Ask your host which multimedia platform is preferred: overheads, computer, blackboard, etc. Ask about the length of time for the talk. If the talk is for one hour, prepare a 40–45 minute talk. You should try to present new data that were not presented in your first talk because this will impress the audience. If you have other projects that you did not present previously, and they have gone well, present them in this format as well. Because in this setting Dr. Kneematoad will get his opportunity to ask all the questions he can, it will be important to state at the beginning the importance and novelty of what you do and what you want to work on when you take your position and have your own lab. It will also be important to be able to improvise on the fly if you run out of time. As before, leave your watch on the podium or next to your notes so that you can keep track of the time without making it obvious. Future goals and approaches should be organized in a format that is similar to the specific aims of a grant application. As you will read in the chapter about grants (*Chapter 3, Gettin' Money*), it is important to have questions/hypotheses in your future research goals that are easily addressed with your system and are of fundamental importance, as well as to have goals that may be more risky. Be careful of proposing too many "fishing trips" in your future goals. A fishing trip is a series of experiments to find new things to study (see Chapter 3). These informal talks can be very informative and you may actually get good feedback and ideas from the group.

> ▪ As in your first talk, be energetic and excited about your work. Do not dwell on what did not work and be sure to emphasize the future importance of your research.

During this visit you may travel around with a realtor to look at housing. Aside from the size of the house or apartment, there are lots of issues that affect housing costs and affect where you decide to live. Your colleagues will be able to tell you about school districts, neighborhood safety, and commuting times to and from work.

NEGOTIATING YOUR POSITION

It's not personal, it's business. Your goal is to get what you need so that you will be successful in your research program. Here again research, teaching, and service are the three parameters that will determine your success. While thinking about how you will approach your negotiation, you should try to get a feel for what the chair's position is. The chair's job is to hire the best person for the position; however, depending on the financial resources available, the chair may be limited in what she can and will do. The chair may need to spread the resources out over several positions. *The Job Comparison Worksheet* (**Appendix 1-7**) will help you summarize your offer.

Our *Next Wave* article ***Greedy Gus: Negotiating an Assistant Professorship*** (reproduced in *The Extras Section*) discusses some of the issues addressed below in more depth as well as approaches to consider when negotiating.

> ▪ The key to your negotiating stance is to hold firm on issues that will be critical in determining the success of your research program, and to be flexible on issues that do not.

Salary

Amount

Salary is always an issue and making more money rather than less is always better. However, the total dollar amount should NOT be your major issue (unless the offer is very low). Average salaries for professors at all ranks and types of U.S. medical schools are surveyed each year by the American Association of Medical Schools and published as the AAMC Annual Salary Survey. This book, which is usually kept locked up in your institution's Dean's office, will give you an idea of what you should expect. The American Association for the Advancement of Science (AAAS) has a website — http://php.aaas.org/membership/salary/salary.php — that allows you to query a large database on faculty salaries based on discipline, institution type, degree, etc. Each of these resources provides a range of salaries based on whether the individual is in the 25th, 50th, or 75th percentile of their peer group. As a brand spanking new Assistant Professor, you are more likely in the 25th percentile than in the 75th. Local cost of living issues will augment or diminish the amount. While all these resources will tell you what scientists earn, the data are at least one year old. The best way to know what you are worth is to get multiple offers. Remember that salary is almost always negotiable and the data sources above provide you with what is within reason.

Salary Source

After the amount, the second most important issue with regard to salary is who pays it. Each institution has different rules. These rules can even vary within an institution's divisions and within departments. Most positions require some form of salary recovery. Salary recovery can come from extramural sources such as research grants or from the income generated from seeing patients. For example, the department may have a policy that it is responsible for 50% of your salary, and you have to generate the remainder of your negotiated salary through your percent effort on extramural research grants. Another example is that you only get a salary while the undergraduate programs are in session, but you may generate an additional summer salary through research grants or by teaching courses over the summer. You also may find that the department requires that you generate all of your salary from research grants. If you are in a clinical department and the job that you have has a clinical responsibility (i.e., you will see patients), then the source(s) of your income may be an important issue (see Clinical Positions below). Thus, there are four issues with regard to salary recovery:

- ▪ How much of my salary am I required to generate?
- ▪ By when do I have to generate this amount?
- ▪ What happens to my salary if I generate more?
- ▪ What happens if I generate less?

The amount of your salary that you have to generate through research grants and/or service will provide you with a goal. If you have to get 100% of your salary you will likely have to have two or more grants to cover your salary and still have money left to do the proposed science. If the goal is between 30 and 50%, then this could allow you to get by on only one grant. If you obtain a grant early or come to the department with a grant that contributes to your salary, then you may be able to negotiate to get some of that money returned to your lab for supplies or to hire additional technical support. If your percent effort on grants exceeds the amount that you are required to get, there will be a surplus for the department. Some departments have procedures in which they "supplement" the labs based on the salary recovery of their faculty. Other departments allow you to "buy out" of your teaching time. Thus, if you cover your entire salary on your research grants, then it is not appropriate for you to be asked by your chair to teach a major course, because your time should be totally devoted to your research.

The "what if I get less?" question is always difficult, and the true answer will not be known until the issue arises. Typically, nothing will happen immediately, but you may not receive a full share of the standard cost-of-living, annual increase that everyone else receives. However, if the department's resources are strained it may mean a salary reduction for you. Thus, knowing how the department deals with this issue up front may make a big difference to you, your overall security, and whether this is the right job for you.

Rank and Promotion

This should be spelled out clearly from the beginning. If the position is for a tenure-track Assistant Professor, you should know the timeline for promotion and tenure. If "tenure track" is the wrong track for you, you will ultimately be miserable. If you are coming into a position from another institution, and you have already established your lab, have funding, and have been productive, you may want to negotiate starting at a more senior level or request that you come up early for promotion. If you come up really early, that is, two to three years early, and you have published several papers or gotten additional research grants, then your past achievements are likely to be considered in the promotion decision. If you come up later but ahead of the normal timeline for promotion, then only what you've done at the current institution may be considered. In this case, you are better off not going through the process unless you have achieved the same level of accomplishment as others in your department/institution who were successful at attaining promotion. Your future chair should be able to advise you appropriately.

Percent Effort

Because most academic positions require that the faculty function in the operations of the department, you will need to know what is expected of you and how much time you will be able to devote to your research program. It is critical that you have at least 50% of your time completely and utterly devoted to research. The three examples of an undergraduate department, clinical department, and basic science department are discussed below.

Research and Undergraduate Teaching

A department with an undergraduate major will likely be responsible for a plethora of courses. The key issue to you is how many courses do you have to teach each year, when do you have to start teaching, and is there any way to reduce the amount? As discussed in

Chapter 7, Being a Teacher, preparation for a class takes enormous amounts of time for the experienced lecturer. Preparing for your classes for the first time will occupy all of your free time. Thus, if you are asked to teach one course in each of the two semesters of your first year, you will have difficulty in organizing and writing your first real grant and in starting up your research program. However, if you like the idea of teaching and interacting with undergraduates, and you have the time (in years) to establish your research program, this position may be perfect.

Research and Clinical Service

Clinical departments in medical schools must provide some form of clinical service. Such services include patient care and diagnostic laboratories. Many faculty are required to fully support the clinical endeavors of departments. As above, the question to the physician scientist is: "how much time will I get for my research program?" A common split is 70–80% research, 30–20% service. However, the definition of what constitutes research time is often fuzzy. Be sure that your future chair, the other faculty members, and you are on the same page with regard to your effort. Talk with the faculty and find out how much time they actually spend on their service component. Your salary is also likely to be linked to the amount of clinical service you perform.

You also need to consider that if you spend only a small amount of time a year performing your clinical duties, you may lose some of your clinical skills and it may be difficult to keep up with the latest and greatest techniques and protocols. Additionally, it will take a longer period of time before your colleagues accept and fully trust your judgment.

Another key to success in the clinical service arena is the ability to have help with your duties. Will there be strong secretarial support? Are there physicians with whom you will partner and share/split your patient responsibilities?

Research and Basic Science Departments

Basic science departments in medical and other professional schools were formed to teach science to their professional students (medical, dental, nursing, etc.). In research-centric institutions, these departments have as their basis a strong commitment to science and most of the faculty spend most of their effort on their research programs. These departments have two general responsibilities that may detract from your research. The first responsibility is the professional courses. Because the number of courses is usually small, most of the classes are team taught with each faculty member being responsible for a small number of classes. Thus, you need to find out how many lectures you will be giving in such classes and when you have to begin.

A second responsibility may be in the organization and management of a graduate program. New faculty are rarely asked to manage such programs. Because graduate programs require faculty to teach courses and serve on thesis committees, exam, and admission committees, etc., you should find out what courses and other assignments you will have to participate in and when you will have to start.

Within these different departmental structures and responsibilities it is common to be able to negotiate lower service/teaching loads over the first two years so that you can get your first grant funded and your laboratory humming. After all, your success is their goal too.

> • Your ability to find the balance that allows you to perform your teaching, clinical duties, or other services, and perform your research is the key to this aspect of the negotiation process. Look at how others in the department have balanced their time. If the department is unwilling to give you what you think you need in terms of research time, then you will ultimately be unhappy, even if they offer to pay you the big bucks.

Research Resources

Next to percent time, this is the most important aspect of your negotiations because your success depends on getting the correct start-up package. You will need to get enough resources (money, staff, and equipment) to establish and run your program fully for two years. This is how much time that most faculty and administrators think that you will need to get through three rounds of study section review at the NIH (see *Chapter 3*). The worksheet in **Appendix 3-3** will provide you with an Excel spreadsheet to work out some of the details. In cases where you believe you need special consideration, you should be prepared to provide scientific justification. "I want it 'cause it's better" will not fly. Here, your research resources should be broken down into the following categories: equipment, space, technical support, supplies, and office.

Major Equipment

These are expensive >$5,000 items that you will buy once (like a biosafety cabinet). If you require a very expensive piece of equipment, be sure to make it clear that you need this equipment for your work. It is possible that the chair may be able to procure other resources to offset her costs for setting you up if the equipment is special.

Minor Equipment

These should cost <$5,000 and are not designed to last forever. These will include water baths, refrigerators, –20° freezers, small incubators, vortex mixers, etc. The best way to get a handle on these items is to walk around the lab you're in now and jot down everything that you use in a month's time and decide if you need this equipment in your lab or if you could share it with a group of labs.

Computers fall into this category as well. While not really disposable, computers need to be replaced every 3–4 years as operating systems and software upgrades no longer run efficiently on the older machines. You will need a computer system for your office, and at least one for every two people you anticipate working for you. You should also consider having a printer and a scanner in the lab as well.

Space

An entry-level assistant professor should expect entry-level space typical for that department. Generally, 600–1000 square feet is offered. The arrangement of the space and the general layout for locating equipment should be compatible with the experimental designs that you employ. If your expensive equipment has special space requirements, this too will have to be discussed.

Some departments have only shared space. In this scenario, faculty are assigned benches and places to put their equipment within a large common lab room. If this is the situation that you are considering, then knowing who your neighbors are may be important to your happiness and success. The advantages to this arrangement are the close intradepartmental interactions between lab workers and the ability to gain more space as your funding increases. The disadvantages to this system are: (1) that your space could contract at some point due to loss in funding; and (2) the potential for personal and policy conflicts between the different labs groups.

Technical Support

It is common to request financial support for a technician or postdoc in your start-up package. The length of time that this person will be paid by the department is negotiable.

Supplies

You will be surprised at the cost when you place orders for everything you need. Initially, you will need about $30–35,000 for supplies for a molecular or cell biology lab. If your research is supply intensive, you should expect to spend $1,000 to 1,500 per person, per month. Ouch! If you use animals, estimate the number that you need per year and the daily costs involved. Each institution will have different costs (per diem) associated with the care of research animals. No matter what the institutional policy is, animal research costs are expensive, so plan carefully.

Office

General office furniture is usually already in place, but if it is not you should allow $5,500 for desks, tables, chairs, and file cabinets. An office computer and printer are a necessity.

Moving Expenses

It is reasonable to expect that your relocation/moving expenses will be covered by the department. This can happen in a number of ways. You may get 1–2 months of your salary ahead of time to cover your costs or the department will just cut you a check for the amount. If you have a research track or instructor position already and have lab equipment, these expenses should cover moving your laboratory. There are certain tax laws associated with personal moving expenses that you may want to brush up on so that you can save money on April 15th. Some institutions will offer housing subsidies or arrangements if the cost of housing is too expensive

Indirect Costs

The topic of indirect costs may come up during your interviews as a magical source of money that moves into the department. What are indirect costs? Indirect costs are collected by institutions to administer grants. These funds are obtained from granting agencies to defray the cost of providing a research environment so that the work can get done. Depending on the institution and its prearranged agreement with the various granting agencies, indirect costs are calculated on the funds that you get to spend on your research (Direct Costs). Indirect costs may be 85% of the direct costs. So, you ask, what do these indirect costs pay for? It isn't clear that anyone really knows for sure. They are supposed to pay for space depreciation, utilities, library collections, administrators to manage your grants and research policies, office supplies, and maybe the college team. Some institutions provide incentives to researchers or departments by providing funds to the department based on the amount of "indirect costs" that the department generates. These funds are typically unrestricted and allow you to hire secretaries or purchase office items no longer allowed by certain granting agencies. Such discretionary or "slush funds" can also provide the freedom to initiate your own seed project and explore new areas of research. Therefore, you should find out about how your future research home handles its indirect costs and what the "rewards" are for being a successful scientist.

NEGOTIATING THE OFFER

Get your offer in writing! The offer letter is your initial point of negotiation. If the chair doesn't offer you a letter, ask when you should expect it instead of demanding it. If the chair says, "Don't worry, my word is solid, just ask anyone," you should worry. It's business. You can always say that your advisor thinks that you should have a formal letter. When you get your letter it may be perfect. Congratulations! If the letter is not up to your expectations and you still want the job, go through each section carefully and decide what is most important to you:

- Start-up funds
- Teaching requirements
- Salary, salary recovery
- Service time, etc.

Prioritize the list. Decide what is critical. In case you need more money for the fancy Sort-o-Meter instrument, you will need to give a clear "science" rationale: "I need it to do the sortometry that I have been doing for the last two years" is usually an acceptable response. If you have multiple offers, then you can use these offers to judge what you are worth, and while each may have different advantages, one of the institutions is where you will want to go. You may use the other offer as a bargaining chip to gain a little more, particularly in terms of salary, salary recovery, and service and teaching time. If you play one institution against another, you may end up at the one you do not want, or nowhere at all, so be careful, polite, and professional in all your dealings. It may even take several rounds or conversations. Remember, no one likes Greedy Gus. Use the worksheets that you have made in this chapter as they will keep you organized and on track during the process. The key is to get what you need. It's Business.

NEED TWO JOBS?

That is, one job for you and one for your spouse. If your spouse is also a scientist or an academic, it is often possible to find positions for both of you either at the same institution or within the same city. While some suggest waiting until you have an offer or a second visit to bring this issue up, you are likely to be asked about your spouse or significant other's career path early in the interview process. Answer honestly. You really don't want to be separated, so you need to let the departmental chair know that she may have to help your spouse find a job. If the department that you are considering is also an appropriate home for your spouse, this will give the committee time to consider your spouse's application. If it is not an appropriate home, informing the chair early will give her time to make arrangements with other departments to consider your spouse's application. If your spouse is not a scientist, it is possible that the university has some influence in the local community and that interview possibilities for your spouse may become available given enough advanced warning.

> • The bottom line is to inform your future institution of your spouse's employment needs so that if they want you, and you want them, everyone — even your spouse — can be happy.

CHAPTER 2

GETTIN' STARTED

"Congratulations and good luck," everyone says as you leave your postdoctoral position behind and head out to be a big-time Principal Investigator (PI). You pack your bags; you show up. Now what do you do? Research of course! Most academic positions have a little more in the job description than just the "science," and therefore you need to be able to decide how you will tackle the academic environment too. This chapter will address issues that will arise during the first three to six months of starting your new position, including:

> - When to start
> - Setting up your lab
> - Setting up your office
> - Getting ready for your role as an academician
> - Time Management

WHEN TO START

It seems as though you should start work as soon as possible after you get the job offer. However, this may not give you the most flexibility in managing your career. For example, in some institutions, the promotion and tenure clock starts on a given date, such that all faculty hired before or on September 1 will start the tenure clock on that date. In this type of institution, if you start on September 2nd or later, then your first year doesn't begin until the *following* September 1. This gains you an additional year on the tenure cycle! This issue may increase in importance to you as funding becomes tighter. If your position has significant teaching and/or service responsibilities, a longer promotion time line may make the difference for success. You should discuss this issue with your new departmental chair prior to agreeing to a start date.

If your institution has a flexible promotion and tenure clock (that is, the tenure clock starts when you start the job), your choice of start date could be influenced by your probable teaching load and access to incoming graduate students. Coming in the middle of a semester may exempt you from having to teach in a course and, therefore, buy you more time to set up your lab. Arriving in the fall semester has the advantage of allowing you to be introduced as the NKOTB (New Kid On The Block) to the fresh crop of prospective graduate students that you might be aching to recruit into your laboratory. If you arrive in April or May, the chances of getting a new graduate student for that year will be reduced because the incoming class of students may have already settled into other labs.

> ▪ Because in many institutions you can come up for promotion or tenure early, having an extra year or even a few months may make the difference in your promotion/tenure decision if you need it. The key is to get as much time as possible so that your tenure package can be as strong as it can be.

A major consideration may be where you are in completing a particular project. If your research is cruising along and the data are being generated quickly, you may decide that moving now will result in a six-month delay before you can complete the project. But if you stay where you are, it will take only a month. Staying for another month or two could also provide you with the critical preliminary data for your new grant application. However, two months may lead to three and three to four ... Before you know it, you've stayed six months and you marched up a dead end or the results weren't what you thought they would be.

To make this decision, consider the difficulty and time required in setting up the specific assays that you perform, the position of your competitors on this project, and how much effort you will be able to put into getting your first experiment completed in your new lab. If you can order equipment and supplies before you arrive, this may save you some down time. Remember, you have to arrange it all yourself.

Of course, family issues may need to take precedence. For example, if you have school-aged children, you may want to begin your new post in time to allow them to get settled in their schools. You may therefore consider moving in August and starting on September 2nd, if the tenure clock rule is a concern. The worksheet in **Appendix 2-1** is designed to help you coordinate starting date issues, as well as other "setup" issues that are discussed below.

Now, wasn't that easy?

SETTING UP YOUR LAB

"Congratulations and welcome," everyone says as your arrive at your new position. The conversation will also include, "Where are you living? Did you buy or rent? How long is your commute, and so on, and so on ...?" After all this is said, the most likely next question will be, "So, whatcha gonna work on?" Certainly, you answered this question in your interviews and are ready with your 54-character title for your new NIH RO1. This is without a doubt Priority One, and if you haven't planned out your first grant, refer to *Chapter 3* to read about the grants process. If you already have a grant, congratulations, you are ready to roll.

Space

Wide open spaces! That's what all new labs look like. They have clean walls, floors, and benches, with only a few dust balls underneath them. New labs also come with a small echo that will be reduced to a hum as the space gets filled with people and the equipment begins to churn. Someone probably said, "Oh, you can get at least 8 people in this 800 sq. ft. space." "Yep," you replied. Maybe you could get them in — they just would have to work in three shifts. Therefore, you need to maximize the use of your space. This statement is true whether you have your own dedicated space or you share space with a group of investigators.

Different types of science obviously require different setups and space utilization plans. Unless you have the opportunity to design your own space, your lab will come with permanently fixed items (e.g., benches, hoods, sinks, etc.) and movable items (e.g., chemical storage shelves, tables, desks, file cabinets, and all your large equipment). The general design parameters of your lab are likely to be similar to others in the department or building that you are in. Before you place a piece of equipment in your new space, look around at other labs in the department, because you may find that someone else has done a lot of thinking about where to put things. They also may have thought very hard and messed it up, too. Now is your chance to observe, ask questions, and learn. It will also flatter people if you ask to see how they arranged their labs, and a certain amount of flattery of your colleagues can be good.

Now that you have looked around and asked questions, you need to make some decisions. Get a floor plan of your lab. This will give you the positioning of the non-movable objects, like benches and hoods, as well as the physical proportions of the space. If you know the dimensions of the equipment that you will purchase, make a reduced-scaled footprint of the equipment out of sticky notes. You should do this with desks, file cabinets, movable shelving, etc. Now you can move things around on paper without throwing your back out by moving the actual equipment. (You will get opportunities to do that later.) If a piece of equipment "just fits" according to the scale of the floor plan you will still need to measure the space yourself, because construction of laboratories and buildings is a qualitative science, not quantitative. Ideally, equipment that will be used for a single assay or preparation should be as close in proximity to each other as possible. For example, placing a centrifuge near a sink so that supernatants can be easily decanted would make using the instrument efficient. Noisy equipment should be in a separate room from the main lab, if this is an option. No one likes the sonicator next to his desk.

There are two common physical layouts and several general strategies to organize the space. The two physical layouts can be grouped into labs with "islands" and those with "peninsulas." Island labs have center benches (the island) and benches against some of the walls. This provides very flexible space that can be used by most scientists. The major disadvantage is that student/staff desks are not adjacent to their workspaces. Peninsula labs have rows of benches that are perpendicular to a wall with a desk juxtaposed at one end. The advantage to this arrangement is that space for single individuals is predetermined and typically larger per person than "island" space. The disadvantage is that this arrangement may make it more difficult to double up your personnel on a peninsula because one person will not have a desk. This could prove to be a problem if your lab funds and the number of personnel grow faster than your space. But that would be a good problem to have.

Lab design can affect your space organizational choices. While there are lots of ways to organize your space, two modes are common: assay and individual. In the assay mode, your lab is organized around specific assays/data collection equipment or workstations. In this mode, you anticipate that everyone will share most of the space and that assigned individual space will be minimal. Here the center island arrangement works well. In the individual mode, lab benches are assigned to each lab member and the assigned space is typically not shared. This setup presumes that each individual will have most of the equipment that they need to carry out their work. This is a common setup for labs that use the tools of molecular biology as their predominant assays. This mode is best suited to the peninsula design. If you choose the individual mode, you may consider restricting the use of radioisotopes and other hazardous reagents to one area of the lab. This will help restrict contamination problems when your new graduate student with the curly red hair splashes radioactive phosphate on the floor. Naturally, combining the two modes may work best if your lab has a limited number of routine assays.

Purchasing Equipment

You only get to do this on a very large scale once (unless you change jobs), so make a list and check it twice. Salespeople like the NKOTB better than anyone on the planet. The commissions from your extensive equipment orders stand to earn them a down payment on a beach house. *They will negotiate!* If you read Chapter 1, you should have negotiated with your chair to get enough money to purchase the equipment that you need to set up your lab. Now you have to make the decisions about the quality of the equipment: will you order the Rolls Royce, the BMW, or the VW bug? Unfortunately, *Consumer Reports* does not evaluate scientific equipment. There are, however, Internet news groups that focus on various scientific technologies that can provide you with feedback about reliability and ease of use. It is also likely that you have had experience with one brand or another while you were a postdoc. In deciding which to buy, you should consider the following list:

- How long do you plan to use the instrument?

> The longer you anticipate using this equipment, the more you may want to invest in a brand that has highly reliable components.

- Will the instrument be the mainstay of a lab assay?

  As with the above, if the instrument will get heavy use, the one with the sturdiest parts is likely to last longer.

- How quickly is the technology developing?

  If the technology is changing rapidly, you may have to replace the instrument in a few years, so a cheaper instrument may be better now.

- Do you really "need" the superfluous features in the more expensive instrument that are not available in the cheaper version?

  Cheaper instruments may not have all the bells and whistles. Sometimes the whistles sound good, other times they are never heard.

- Have other labs in the department chosen one brand of instrument over another?

  The feedback of the owners will be valuable. Also, if the equipment has accessories, this may allow you to share them with your neighbor (e.g., centrifuge rotors) if he has the same basic instrument.

- Does the local service available for different instruments vary?

  Say that everyone in your department has a DuPlex Inc. GammaZoid. The instrument requires regular calibration. The DuPlex service person has an office in town. Such factors may lead you to buy the DuPlex. However, you may discover that the TriPlex Inc. instrument does not require regular service, has some added features, costs less, but lacks the name recognition of DuPlex. In this case, choosing the TriPlex instrument maybe the way to go.

  - The point is, consider the servicing of your new equipment before you buy it.

- Footprint?

  If many of the factors noted above are equal, fitting the instrument into your space may be a primary determinant.

- Price?

  Buying the DuPlex may result in your not being able to buy additional equipment with your start-up funding. Thus, choosing the TriPlex may allow you to get more "bangs-for-your-bucks."

- Does the more expensive instrument look "cooler" or have more buttons?

This is self-explanatory.

If you are still undecided after answering these questions and if the equipment is small enough, vendors will often let you test-drive a demo instrument in your lab for a few weeks. If the equipment is not portable and you have never used it in the past, they may fly you to a lab that has one for you to try. This is a great way to decide if the Triplex GammaZoid will do or if you really need Duplex's model.

Now that you have a headache from thinking about each of these questions, you should be able to make an informed decision. If your objective is to hurry up and equip your lab without thinking through these variables (that is, avoiding headaches), then you will probably regret it later. No kidding.

The worksheet in **Appendix 2-2** is designed to help you list and sort through your equipment purchases.

Purchasing Supplies

With the exception of salaries, ongoing supplies expense will be the most costly part of your operation. Using your established budget (see *Chapter 4, Managing Your Money*), buy the items that you need now. During your initial start-up phase, the salespeople may offer you big discounts on items if you spend big $$$$ with them. It is OK to play different general supply companies against each other. They live for the competition. The key here is to get what you need. While salespeople may often show up at the wrong time to talk to you, you will find that being consistently courteous to them may get you a good deal or two now and throughout your career. The more you order and the longer you buy from them, the better the deals you will get. To avoid having salespeople interrupt you while writing a grant application or doing experiments, ask them to call ahead to set up appointments.

The worksheet in **Appendix 2-3** is designed to help you organize your initial supply orders.

SETTING UP YOUR OFFICE

Small, confining space! In fact, it will seem very small, and the view, if you are lucky enough to have a window, may not be what you expected. Sorry, this situation will not change for a while. The key now is to make it livable and workable. As an academic scientist, you will spend a lot of time in this space. Good lighting, comfortable chairs, and peace are key issues. If you have the decision over the selection of office furniture, look at what other faculty have (sounds familiar, doesn't it?). Buy a good chair. This is where you will place your most important asset (you). The furniture that you buy will be with you for a long time, so choose wisely.

There are some key items that you need to consider: desk, chairs, table, guest chair or couch, computer system, file cabinet, book shelves, answering machine/voicemail service, coffee maker, and small refrigerator. It's impossible to get all of these items into a

100 sq ft office, so don't try. Couches, chairs, tables, and desks all fall into the general category of furniture that you will use on a daily basis to meet with people, discuss science, read, and do your writing. If you have the space, a small table with a few chairs can provide a comfortable setting to discuss data with your students and visitors. A couch, besides being useful for napping, also provides a relaxed environment to discuss your work with visitors.

Now that we scientists do much of our own secretarial work, desks come in numerous configurations to allow computers to be used. Find one that fits your body and has the drawers and accessories that you need. You will need to have a keyboard at the right height for computer use, as wrist injuries from poor ergonomics are not uncommon. This is very important, so if there are examples around to try out, take them for a test drive. Make sure that your chair is adjustable so that you are at a comfortable height at your desk/keyboard.

The Office Computer Is Your Friend

This plastic contraption will serve as the primary instrument of your creativity, brilliance, frustration (those angry letters that you never send), and communications with the outside world. The simplest advice is to get the *best* you can afford. However, you need to remember that computers, like 1970s American cars, are designed to run for 3–4 years. At that point, the software you use will require more space and memory than your computer has and it will be time to upgrade — everything. This being said, you still should not buy the cheapest model. You will need to decide between the two major platforms, Mac vs. PC. There are arguments for both, and people have very strong opinions about this topic. You may have a preference as well. The important issue here is to buy the right configuration. Get enough memory so that you can use all the programs that you need efficiently. If your research requires image processing, you will need more RAM than if you just use graphs. These days, hard drives are huge and it's hard to imagine that your data can fill 200 GB of space to capacity. (Of course, this is also what everyone thought in 1988 when hard drives were 10 Mb.) Flat panel monitors are recommended to save on desk space. We also recommend that you buy as large a monitor as you can afford. The ability to have multiple documents open at a time will come in very handy when writing grants and manuscripts.

You will also need some sort of backup system. This will come into play the week that your grant is due and your computer has completely crashed, and you are looking at the earliest version of your grant from 10 drafts back. You may think that this does not happen, but it happens to almost everyone. Computers "know" when you are under the gun, and they revolt at the most critical time. Writable CD/DVD drives present an inexpensive, stable backup system where most of your important documents can be stored. Flash drives and other removable devices are great for routine backups but are not permanent and the data on these can be lost. Fancy tape drive devices or simple external drives can be programmed to backup all your lab computers on a daily basis. Such systems provide the greatest safety net for your work and data and require little effort to maintain once they are set up. Some departments provide a backup service to their server. While this is great, faculty often become complacent about their backups, only to discover that the drive or tape was full for the last 3 weeks before their computer crashed and burned.

> ▪ The key here is to make sure that YOU can and DO back up your work.

Computers for Your Staff

You will need at least one or two computers in the lab. It will need to have Internet access for database searches, etc. Depending on your type of research, you may need many computers. Unless the data generated by your staff require serious number crunching or image processing, you may be able to save some money in this area. A slower processor, a smaller screen, or a little less RAM will not hinder the progress of your laboratory. If all of your data are generated through the use of the computer, this is where you should put your money. The larger your staff, the more computers you will need.

Scanners and digital cameras coupled with a photographic quality printer have for many supplanted the use of professional photographic services for creating quality figures for publication, posters, and presentations. On-line manuscript submissions for most journals have all but eliminated the need for high quality photographic systems for routine work. However, if your work requires hard-copy photographs, you will need to make sure that you have access to a photographic quality printer. Photographic quality printers are costly items and are easily shared among a large number of laboratories. If your department does not have one, you should be able to find one somewhere on campus or convince the rest of the faculty that the department needs one so your chair will buy it with departmental funds.

> ▪ In setting up your new lab and office, use common sense to make decisions about organization and what to buy. However, don't dwell on the issues for long. Make your decisions and move forward. The goal is to get your lab and office operational.

GETTING READY FOR YOUR ROLE AS AN ACADEMICIAN

In addition to your lab and office, you will need to get ready to begin your life as a member of the faculty of your institution. Faculty are the heart and soul of a university — at least, that's what we faculty think. In addition to research, there are other faculty duties/responsibilities: teaching, service, and administration. As a beginning faculty member, your major responsibility aside from your research is likely to be your teaching. If your position has a service component, then this will also be a major responsibility. Few junior faculty have large administrative burdens, so this is a minor concern, at least at the beginning of your career.

Because all of these duties will distract you from your research, it is important to have a plan. The plan should start with the university calendar. Since the beginning and ending of classes are important events in the university, you should mark these dates on YOUR calendar. These dates are significant because they determine when classes will be taught, when the students will be around (undergraduate, medical), when the seminar programs will be, and typically when there are faculty meetings. Thus, the majority of the time between classes, like the summer break, can be almost completely devoted to your research.

Teaching

Hopefully, you arranged to have a time lag before you have to prepare your lectures and teach your classes. If this is the case, you should find out what classes you will be expected to teach when the time comes. This will achieve two things. First, it will (hopefully) avoid the last minute issue of: "Oh, why don't you also teach Dr. Brillodooz's lectures?" The second point is that it will give you time to think about your lectures, and, if other faculty members are giving them this year, you will have an opportunity to sit in on the lectures and find out what they teach. This latter point will save you enormous time next year when you take on the lecture topic yourself. Did we say enormous time? Yes we did!

If you have to teach right away, you should refer to *Chapter 7* of this book to get started planning your lectures.

Service

"Great! Now we have someone new to do all the work." Service, as will be discussed in *Chapter 9*, has many aspects and definitions. Service responsibilities can include clinical service for physician scientists, running a diagnostic or core laboratory, committee assignments, student advisement, and administration. As is the case with teaching, you may not be asked to do anything right away. If this is the case, your idleness can give you opportunities to learn. Look at what others in your department are doing to help it run. Some service jobs may be more in line with your personality, skills, and ambitions. These are the jobs that you may want to do when you are more established. Find out about them. "Showing an interest" is the best way to be considered the heir apparent for a service assignment you may want. You will eventually be asked to do something, so be alert to the jobs that exist. It's better to do something you want to do than to do something you don't, and you could get stuck with the latter.

If your position has a service requirement, such as seeing patients, running a core facility, coordinating admissions, etc., be sure you know what is expected on Day One or preferably when you negotiate your position. These issues can include the number of days/weeks/months that you must do this service. Be sure that your chair, and your colleagues have the same idea that you do about the time you must spend in the service. If the service occurs once a year (e.g., admissions) find out the timetable.

> ▪ If you will be responsible for a core lab or a regular service facility, establish rules for your time, what you will do, and what you won't.
> ▪ If consultation with the users of the core or service is necessary, establish regular consultation times each week. This may be difficult at first, but once it is in place most people will go by your schedule.

Most faculty spend a good deal of their time serving on a variety of committees. Knowing what is expected of each committee is important to how you will participate and how much time you will need to devote to its success. *Chapter 9* will discuss ways to shorten the time of committee meetings and achieve your desired outcome. While biomedical research scientists rarely get promoted or earn tenure based on their committee work and institutional service, it is expected that you perform these aspects of your job well. Time management is the key to being able to allocate time for service as well as for your research.

TIME MANAGEMENT

As you prepare your "to-do" list, you should attempt to a establish a set of time management rules. A key to time management is to avoid duplicating your effort. This will give you more time for everything else. For example, you receive a note from your chair to send your CV to the head of another department for a joint appointment there. If you now put the note on a stack of to-do items, you will read it several times before you act upon it. But, if you print and send your CV (which you created in *Chapter 1*) and give it to the chair, you can throw the note away and the task is complete. We suggest the latter. Prioritize your tasks. Tasks can be classified into four general categories that should help you manage your time:

> - Imperative and vital
> - Not so imperative but vital
> - Imperative and not so vital
> - Neither imperative nor vital

Effective time management dictates that you deal with all tasks that are imperative first, followed by those that are vital next. Vital in this case refers to your career. Tasks that are not imperative or immediately vital should be performed as a break in the routine and to relax. Here are some examples.

Imperative and Vital — All items that impact your career/life that have deadlines! Gotta do and Gotta do now!

- Your NIH grant Program Director needs you to email her your Other Support pages by this afternoon so that she can try to squeeze your grant into this year's pay cycle.
- Your grant application is due today.
- An abstract to a meeting is due.
- Your completed tenure package is required by your chair.
- Return call to your spouse about dinner.
- All items for which the deadline was yesterday!

Vital but not Imperative — Most lab management, teaching, and service. Gotta do, but it can wait a little.

- Going over experiments and data with lab staff.
- Preparing your teaching assignments and teaching class.
- Writing your research papers (this can become imperative depending on your competition in the field).
- Filling out the required paperwork so that your lab can use radioactivity or work with animals (unless there is a deadline for this).
- Grading papers/exams (unless they are overdue!).
- Keeping up with the literature (that is, reading).
- Discussing progress of your students with other faculty members.

Imperative but not necessarily vital — This is a difficult category to describe due to the relative value one places on tasks. These are items with a deadline that do not impact your career immediately. Many non-science life issues fall into the imperative category. Sometimes items that were neither imperative nor vital end up becoming imperative and vital as the deadline for their completion looms closer. Gotta do it sometime, best to get it done now instead of waiting until tomorrow.

- Reviewing a manuscript for a journal.
- Reviewing a colleague's manuscript or grant application.
- Service responsibilities (not life threatening) that you have put off until the deadline.
- Settling a disagreement between two lab members over the selection of the radio station (we will discuss this type of problem in the *Chapter 4*).
- Arranging to talk with the salesperson about a new piece of equipment.
- Getting a piece of equipment fixed that you will need soon, but could do without today.
- Defrosting the lab freezer because you ran out of room.

Not Imperative and not immediately vital — All other tasks! Tomorrow!

- Getting the menu ready for next month's departmental retreat. Items like this one will move up in importance as you put off their completion. It's best that such items never become imperative and vital.
- Throwing a party for your lab.
- Talking with your fellow faculty members about Atlanta Braves baseball.
- Hanging pictures in your office.
- Ordering your season tickets for the Braves.
- Finishing the second edition of this book.

A more detailed discussion of time management issues is provided in our *Next Wave* article **Where'd My Day Go!** (in the *The Extras Section*).

> - You now have made a large number of decisions about how to set up your lab, your office, and career. The most important point is to have reasons for your actions. If you get stuck on an issue, ask your colleagues for their advice. Before you know it, your lab and office will feel like home sweet home.

CHAPTER 3

GETTIN' MONEY

"No way I can participate in that! I've got a June 1 deadline." This scientist is referring to one of the deadlines for grant applications to the National Institutes of Health (NIH), as well as to the fact that normal life is put on hold while preparing one of these applications. The ability to do scientific research depends on getting enough funds to run your lab. Because even in the best of times research funds are limited, there is considerable competition for research dollars. Choosing the right project and funding agency and writing the grant so that others are excited about your ideas are critical to your success as an academic scientist. Accordingly, the specific aims of this chapter are to:

- Discuss the different funding agencies and types of grants
- Discuss grant writing strategies to produce the perfect grant
- Discuss how to prepare the budget
- Present the view of grant reviewer

FUNDING AGENCIES AND TYPES OF GRANTS

In general, all funding sources follow a straightforward series of steps: funding agencies solicit grants; scientists submit applications; the agency receives them and assigns them to review groups; the proposals are evaluated; some are funded; and a critique of the proposal is provided to the applicant. Understanding the process, making the right decisions about how to approach your project, and viewing your ideas from the point of view of the reviewer will propel you on the way to success in obtaining extramural resources.

There are many different sources of funds available for biomedical and biological research: federal funds, private charitable research funds, and local institutional funds. These sources differ based upon the total amount of money that they will provide, as well as the range of research projects that they will fund. For most academic scientists, federal funds are clearly the most important source of research dollars because not only do they provide the most money to you, but also to your institution as indirect costs. (While you may not be thinking about those indirect costs, your chair and dean are!) Private charitable funding agencies support specific scientific fields and endeavors. They usually provide less money for fewer years than federal grants and do not provide the same level of support to your institution.

Federal Funding Agencies

The National Institutes of Health (NIH) and the National Science Foundation (NSF) are the major federal funding agencies in the United States. Other branches of the U.S. government also support biological and biomedical scientific research, including the Department of Defense and the Justice Department. Unlike the NIH and the NSF, the scope of these latter programs is limited. Due to the highly competitive nature of federal grants, your track record in procuring them is one of the many criteria that will be used to evaluate your future potential when you come up for promotion. The sooner you get your first federal grant, the better.

NIH grants support the widest variety of biomedical research. While most people are familiar with the investigator-initiated award, known as an RO1, the NIH offers a wide range of grants. Such grants can provide funds for salary, student training expenses, research supplements, sabbatical support, and institutional equipment purchases. Additionally, the NIH sometimes announces "Requests for Applications" (RFAs) for certain research programs that the government believes need to be targeted. Information about NIH grants can be found at their website, http://www.nih.gov.

The NSF funds basic science research that is NOT health or disease related. NSF funding spans fields ranging from astronomy to zoology. To get a biomedical research grant sponsored by the NSF, you need to focus on the basic science or biology of the system being funded. For example, a research project that investigates DNA repair, which can be linked to some forms of cancer, can be pitched to the NSF through the use of model genetic systems rather than through studying human tumors. NSF grants are excellent sources of funds for those biological researchers whose projects are difficult to relate to specific diseases. If you have a project that you would like to submit to NSF, call the program officer of the appropriate research group at the NSF and talk to him or her about your proposed project to find out if the proposal would be accepted for consideration. The program officer can also give you important information about the logistics of NSF applications, including the due dates for the various programs, as well as the instructions for the electronic submission of these applications. A list of the biological science related funding groups at the NSF can be found on their website, http://www.nsf.gov.

Private Funding Organizations

Charitable organizations collectively provide enormous support for biomedical research. Organizations such as the American Cancer Society, American Heart Association, Multiple Sclerosis Society, and the American Lung Association fund a substantial number of grants for basic scientists at all stages of their careers. A list of some of the major organizations and their websites appears below, along with websites that provide links to many organizations that fund research grants. Some organizations such as the Community of Science allow you to search their websites by keywords related to your research to find funding opportunities. While an organization's name may imply a specific, highly focused area of research, the organization may in fact support basic science research. In addition to reviewing the Internet sources listed below, check with your university's sponsored programs office to find out the ways they can help you locate appropriate funding sources for your research.

Source	Internet Address
Websites for Major Funding Organizations	
American Cancer Society	www.cancer.org
American Heart Association	www.americanheart.org
American Lung Association	www.lungusa.org
March of Dimes	www.modimes.org
Department of Defense	http://cdmrp.army.mil/research.htm
Multiple Sclerosis Society	www.nmss.org
National Institutes of Health	www.nih.gov
National Science Foundation	www.nsf.gov
On-Line Funding Resource Search Programs	
Community of Science	www.fundinggopps2.cos.org
Grants Net	www.grantsnet.org
Illinois Researcher Information Service	www.library.uiuc.edu/iris/
Society of Research Administrators	srainternational.org/cws/sra/resource.htm

Local Research Grants

Most universities have competitive programs for seed grants in support of young investigators' research. The purpose of these grants is to provide the investigator funding to obtain preliminary data that can be used for larger grant applications. These grants are often very small ($10,000–30,000) and realistically provide only supply funding. Because the payout is small, the time you spend writing these grants should be in line with the net result. If you have sufficient start-up funds to generate your initial data and provide proof of the principles of your project, your time will be better spent collecting additional data or writing for federal or private agency funds.

- Don't spend too much time on small local grant applications at the expense of the time you have to spend on extramural grant applications.

STRATEGIES

When should I put in my first grant or for that matter any grant? How much data do I need? How should I organize the project? Is this different from the way I present it in the application? How much money should I ask for? How long does it take to write a grant? These questions are the basis of grantsmanship.

When to Apply?

The first thing to do is to go to the website of the funding agency or organization and *read* the directions. The dates for submission will be indicated, and although some of your colleagues might say: "Deadlines, schmedlines. I always send my applications in late, don't worry about it," you *should* worry about the deadline and be sure to meet it.

You will need preliminary data in your application to: (1) demonstrate that the system or questions that you are interested in addressing will provide novel information; (2) demonstrate the fact that you have the capability to do the experiments that you propose; and (3) characterize any novel reagents or technologies that you describe in your proposed project. **In short, you need to provide the study section/review panel with sufficient proof that you can do what you propose**. If you are able to continue the project that you had as a postdoc, then you should have lots of data and proof that you can do the work. If this is the case, then you are ready to submit an application.

If you are establishing a new system or experimental assay, you will need to show that you are able to establish that system in your lab or show that you can perform the new assay before submitting the application. If you do not have time to develop *all* of the assays, having letters of collaboration from established investigators to state that they will help you or teach you the assays will alleviate this concern. As you become more senior, your lab will have developed the reputation for development of assays and the strict requirements for preliminary data may be somewhat reduced. All investigators must demonstrate the feasibility of their systems to peer reviewers throughout their research careers. So get used to it.

Organizing the Project and the Grant

While few of us like to write grant applications, the process serves to focus our ideas and to organize them into a logical progression of experimental procedures. A well-organized grant proposal serves as a template for your laboratory's research activities for the project period. Because you will be competing for funds with more seasoned investigators, you will need to get "grantsmanship" experience and you will need to understand the system. The very best way to understand the system is to be involved in the grant review process. The *Catch-22* is that if you are not funded or are a new investigator, you do not have the experience to review and prioritize the grants of others and you are therefore unlikely to be asked to participate in the process.

The second best approach to gaining grantsmanship experience is to read your colleagues' grants. Ask your colleagues — both middle level and senior level — for copies of grant applications that were funded. Explain to them that you are trying to develop your style and strategy for your own projects. If your colleagues are kind enough to allow you to read the critiques of their grants as well, that would be ideal. If a grant was funded based on a revision that was submitted, this is even better, because now you will see how your colleague addressed the problems that the reviewers had with the initial application.

Read four or five grants. It will take you about three to four hours to read each NIH RO1 application. As you go through each grant, pay attention to the style and the type of information provided. While the ideas are paramount to getting any grant funded, the presentation of the ideas, their novelty, their significance, and the ability of the experimental design to adapt to the results are crucial to a high priority funding score.

Now you are ready! Well, at least you're ready to get ready.

So what should you write about? As you were leaving your mentor's lab, you probably had a talk with her about which research projects you could "take" and which ones you couldn't. Competing with the old boss is not the best approach and therefore this issue is sometimes sticky. Regardless of your discussions, you probably have a few good ideas to pursue in your first grant application. In writing your first grant applications you should pursue *fundamental* problems of biological and/or medical importance. This will allow you to more easily justify the significance sections of your grants and work, and will provide the study sections and review groups with a reason to fund your grant over others. Targeting your question to specific diseases may lead to opportunities for private funding foundation sources to support your work. Because complex problems are not easily conveyed in the limited number of pages allowed in an application, the more basic/fundamental the research problem, the better. Additionally, fundamental problems affect more areas of study and are therefore more applicable to many fields. This is often a criterion of highly ranked journals: is this appropriate for a broad audience? If your area of expertise is not in a fundamental problem, you will need to relate it to one.

Research problems can be broken down into short- and long-term goals. For example, understanding the mechanism(s) that lead to diabetes would be a long-term goal; whereas characterizing a series of peptides that lead to immune responses to receptors on the surfaces of islet cells would be a short-term goal. Among your short-term goals, some will be high risk, while others should be of the "bread 'n butter" category for your expertise. High-risk projects are needed to hit the home runs in science. However, because high-risk projects often crash 'n burn, you need to design your research around "doable" projects that will allow you to be productive, creative, and move towards your main long-term goals. So, you need both long- and short-term goals and a collection of experimental approaches that lead to their completion. The short-term goals should be mostly low to medium risk with a small number (very small) of high-risk projects.

WRITING THE PERFECT GRANT

How do you know that you've written a perfect grant? It gets funded! OK, that was too easy. Most grant applications follow a format that serves to provide the reviewer with all the information that they need to judge the significance and feasibility of your proposed work. Once you learn how to use the format (successfully, that is), you will find writing grants to be just a normal part of your job as an Academic Scientist. The discussion that follows is based on the NIH format because it includes fundamental elements of all research proposals. The NIH grant application format has four sections: Specific Aims; Background and Significance; Preliminary Data or Progress Report (the former for new applications and the latter for competing renewals); and Experimental Methods and Design. **Appendix 3-1** provides a general outline for organizing an NIH application.

Specific Aims

Enthusiasm for your grant begins with the Abstract and Specific Aims page. Both must highlight the importance of the project, the novelty of the project, the major questions, and how you will approach answering these important questions. In fact, you can write a believable research project about anything. **Appendix 3-2** presents a generic NIH-style Specific Aims page that is written about a nonsense project, but is nevertheless convincing because it covers the required features noted below:

- **Disease or medical relevance of fundamental importance**.
- **History**. This is not necessary, but if there is an extended history to the problem at hand, mentioning it strengthens the significance.
- **Long-term goal(s)**. The long-term goal provides the depth and breath of the project and lets the reviewer know where you are headed.
- **Background of the system**. The background needs to be presented in a very concise manner, such that the specific problems or short-term goals can be understood. If there are discrepancies in the field that you will address, they should be briefly laid out here as well.
- **Hypothesis**. Most funded research projects test hypotheses. While it may be obvious to you that you are testing one, you need to explicitly state what it is. The hypothesis could be broad or very specific and narrow.
- **Questions**. Some investigators like to pose a series of questions. Like hypotheses, questions can be either broad or very focused and essentially describe what the grant is going to be about. Hypotheses and questions presented on this page tell the reviewer what you are going to do, solve, and determine.
- **Novel reagents or approaches**. If you have developed a novel or unique reagent, such as a mutant, antibody, transgenic animal, etc., you should include a brief description of the reagent and how you will use it to answer the questions and what new information you can obtain using it that could not be obtained before. This is also true for any new assay or technology that you have created.
- **Future**. A sentence predicting how your study will aid humanity closes the paragraph.
- **Aims**. Aims can be listed as a single sentence with succinct phrases describing the goals of the project. Other strategies can also be used. Short paragraphs, stating additional details about the approach to be used are added to the aims. In other examples, the hypothesis to be tested in that aim is stated with a short list or description of how the hypothesis will be tested.

As you read through the example in **Appendix 3-2** and your colleagues' grants, diagram the specific aims pages so that you can see how they are put together. You will want to present a concise, clear, description of your project and its significance. By the time the reviewer has read your Specific Aims page, her enthusiasm for your grant needs to be high. So, if the reviewer doesn't understand why you or anyone else would or should do this project after reading the Specific Aims page, then your grant application is in trouble — with that reviewer, anyway.

Background and Significance Sections

"What do you mean, the reviewer will not be an expert in my field?" In most cases, your reviewers will not be expert in your specific field, but they will have knowledge of your area of science. Thus, the Background and Significance section should be written to provide the reviewer with all of the information that she needs to review the grant and get excited about your proposed questions, not to learn ALL about your field. Unless the chronological history of the field is important, present the relevant information to bring the reviewer up to speed, state the problems and why they are problems, and tell the reviewer where in the grant you will be addressing or solving these problems. A separate *significance* subheading will allow you to clearly state why you think your field of study is important and how your proposed experiments will find the missing link. Again, while reading your colleagues' grants look for these items.

Preliminary Data/Progress Report Section

As described above, the point of showing preliminary data is to convince the reviewer that you can do what you propose to do. Showing an ugly western blot, even though it "worked," may not convince the reviewer that you are The One to do this project. Begin this section by summarizing what will be presented in the section. This gives the reviewer a heads up on what to expect. **Appendix 3-1** provides a general outline for this section. For each subsection that contains experimental data:

- provide a title that is informative and conclusive;
- set up the problem;
- describe the experiment;
- address the figure or table;
- describe the controls;
- draw your conclusions;
- address caveats in the interpretation and what also needs to be done or where the data are headed; and
- state where in the grant those experiments will be discussed.

All figures should be labeled as if they were going into a journal article. The reviewer will not be a happy camper if she is spending excessive time trying to figure out what the units are, which band is the one that you think is important, or trying to figure out which clone or plasmid you are using. The figure legend should be succinct and complete. You can provide additional information in the legend about the experiment that was not stated in the text of the grant. The figure size should also be large enough for the reviewer to read without a magnifying glass. If you need to have glossy photos in the grant, they can be pasted in or included in an appendix, but the size of the figure in the grant has to be the same. The reviewers get the "original photocopies" that you provide, allowing you to paste in the fancy photo or to use a fancy color printer for those pages that require color.

Experimental Design and Methods Section

This is the nuts and bolts of the application. A logical progression of experiments addressing the proposed aims is required. In addition, you must convince the reviewer that your experiments will be interpretable and that YOU will be able to interpret them. If the data do not prove your hypothesis, you must be able to redirect the investigation to follow the data. You must also provide alternatives for your experimental design in case your initial approach is not successful. This means that you need to identify the potential pitfalls in your brilliant plan. Following a motif like the one presented below (**Appendix 3-1**) will help you stay on track.

Aim 1 — Restate the Aim

Rationale — provide a brief description of the problem, state the hypothesis(es) to be tested and questions that you will address. If there is a list of questions or hypotheses, be sure to number each one, as they will be your outline and guide through the aim. Again, describe how you will use those novel reagents or assays.

Approach(es) — this is where you provide the design of the experiment. If there was a list of questions or hypotheses, restate them in the form of subtitles. Now describe your experimental procedure. The details that are important are those that are unique to the system that you are using. In each approach or series of assays, be sure to state the purpose of the experiment. Do not forget negative and positive controls.

Alternatives and Interpretation of Data — Incorporate your alternative strategies for each approach. At the end of each series of experiments, you should state how you intend to interpret the data. If the experiments provide you with an if/then scenario, be clear on how you will follow up. If you do not want to follow up on a particular scenario, then stating that this scenario is "out of the scope of the current application" is an easy way out. However, this "out" has consequences if the reviewer believes that such a direction is more interesting than your proposed work. If the follow-up is really required, then you should propose it. Because this may lead the grant in another direction, stating the types of approaches with the appropriate references is often sufficient to convince the reviewer that you will do the right thing.

- Aim Summary — You have just written 2-3 pages of elegant experiments and the reviewer should know exactly what you plan to do, right? To make sure, provide the reviewer with a "sound bite" or two of the aim. This should be a short paragraph that the reviewer can extract from your grant to help her write her review and present it to the committee.

Timelines and Priorities

Here is your opportunity to state how you prioritize the experiments that you propose. This can be written as text, presented in graphics form, or both. Placing the risky experiment at the end of the timeline allows you to attempt to convince the reviewer that you will not be spending your entire funding period on it. It will also tell the reviewer whether the aims will be carried out simultaneously or in sequence. This section should be brief.

> **Survey Says**: A distinguished Full Professor commented — "After placing the first four aims on paper, a first-time grant writer should consider choosing one and develop it into an entire grant."

Tips

If there is a sequence to the aims, you must be careful not to have the entire project depend on completion of Aim 1. If the reviewer does not believe that Aim 1 can be completed, then the entire grant is a wash. Having some aims that are dependent on each other is fine as long as there are others that are independent.

Some investigators place the precise details of their experiments at the back of the grant in a section they label "Methods." While this may make a section less cumbersome, the Methods section is either not read carefully or, if it is needed, the reviewers have to shuttle back and forth in the grant and find the section that they have to read in order to understand your experiment. Such sections are not the best way to present information. If a technique is new, then placing its description in the flow of the grant in a different font will allow the reviewer to skip over it or read through it. This way the information is where it needs to be.

Remember that while your grant is your baby, the reviewer has 8 or more other grant applications to read. The faster the reviewer gets through reading your grant the more enthused she will be. Small-type fonts won't be appreciated, nor will having the text fill the page from the extreme edge of one side to the other. If your grant is longer than the required limits, then you are not concise and clear, or you have proposed too much work and the reviewer will comment that the proposal is unfocused. In addition, a grant application must be spell-checked, carefully proofread, and grammatically correct because mistakes are considered sloppy. Sloppy writing means sloppy science, doesn't it? If English is not your strongest subject, then get professional help from a technical writer or editor. If you have to pay for this type of professional assistance, it will be worth the money.

On the day you have to mail (submit) your grant, expect everything associated with getting your grant into the mailbox to go wrong. Expect the photocopier to break down, your desktop computer to freeze up, and the secretarial staff to be overwhelmed with other faculty members' grants. Be prepared for these problems and others so that you meet the deadline with a complete grant application that has the correct number of copies of the entire grant. Finalize and copy certain parts of the grant like the appendices well ahead of time to avoid as many headaches on grant mailing day as possible. With several agencies moving to electronic submission, it will be important to become proficient with converting your grant into the appropriate file format. You don't want your last day to be spent learning new software. You may also have to get specific passwords and filing numbers for e-submissions so that the funding agency can collate your application with your institution's official signatures, etc.

The Most Important Tip

Ask your colleagues to read your grant before you submit it. Receiving an in-house critique and having the opportunity to fix the problems before your grant is mailed out is much better than having to address the problems in a revised application. Which of your colleagues should you ask? Ask colleagues with diverse scientific backgrounds. As with the agency reviewer you are likely to get, your colleague doesn't have to be in your field. But, if you give your grant to a colleague just two days before it is due, he will not have time to carefully review the proposal nor will you have time to incorporate any meaningful changes he recommends. Your colleagues are busy, so give them a week to review it and yourself 4–5 days to incorporate their suggestions. While it is good that your colleague tells you about your grammatical mistakes, this is not all that you need. You need science and grantsmanship review. Faculty who read your grant who are experts in your field should be able to tell you if the science is sound. Those outside your field may be able to do the same, but they also will tell you what they missed.

When going through your grant with your colleague, **do not** argue with him about his comments. If he didn't "get it," then the reviewer won't "get it" either. Instead of arguing, come up with ways of restating your points or redesigning the section so that both your colleague and the reviewer will understand it. If you have enough time, giving different colleagues different versions of the grant will allow you to test your revisions as you get ready to prepare the final product. Planning for an in-house review means, of course, that you have to prepare a draft at least 3 weeks prior to the agency deadline, so keep that in mind.

> **Survey Says**: Tenured Professor — "... have seasoned successful grant writers review your grants before submission; do not be too ambitious in grant proposals; respond to reviewer's critiques carefully and positively when revising proposals ..."

THE BUDGET

How much money and how many years should I ask for? You should ask for what you need to complete the project. This is what the funding agency believes that you are doing when you fill out the budget pages. The NIH form, page 4, which can be downloaded from their website, http://grants1.nih.gov/grants/funding/phs398/phs398.html, provides a breakdown of the general costs associated with running a lab. An example is provided in **Appendix 3-3**. As discussed in *Chapters 2 and 4*, about budgeting, this form has categories for personnel, consultants, equipment, supplies, travel, patient care, and the most important one, other. Even senior faculty seek help from their departmental business manager in designing their grant budgets. Your departmental business manager will be more than happy to teach you the ropes. Because this takes time, go to him early in your planning process so that you won't have to deal with budget items while you are concentrating on the important issue of what you will write about. Presented below is a discussion of each of the different budgeting items. However, as presented at the end of this section, NIH grant budgets under $250,000/year in direct costs are not broken down into individual categories. While the forms for these so-called modular budgets are different, the planning process is the same (see below).

Budget Categories

Personnel

Unless you have to provide care for patients or large animals, personnel will be your most expensive category. If you are starting out, the study section reviewers and the NIH will expect that you will be spending a substantial amount of your time on this project. Most new investigators devote at least 30–50% of their effort on their first grants. This figure is expressed here and in the budget justification. It is also here that some of your salary is recovered. How many other people should you have on this grant? Most projects require 3–4 full-time professionals. Thus, you should include salary requests for a technician, graduate students, and/or postdoctoral fellows.

Equipment

Now is your chance to ask for that GammaZoid and any other necessary piece of equipment. You will need to justify why you need the requested equipment. The only true justifications are that: (1) the equipment is essential for the project; (2) the instrument will greatly improve productivity; (3) no other one is available to you; and/or (4) it improves the safety of performing the experiments. If you do request equipment (>$5000), get a quote from the manufacturer and attach it to the budget justification.

Supplies

Depending on your research field and protocols, your supply category will be different. The example in **Appendix 3-3** is based on a project that uses animals and performs assays that are cellular or molecular based. Here the amount requested is approximately $10,000–12,000/year/FTE (Full-Time Equivalent employee). Some of the individual categories are shown and are likely to be different for your grant. As noted earlier, seek advice from your colleagues and departmental business manager.

Travel

One meeting/year for the PI and a student are typical requests.

Other

This is the all-encompassing category. Note from the example that items range from publication expenses to service contracts, to animal maintenance costs. Costs to use shared instrumentation are included in this category as well.

How Many Years Should I Ask For?

The best advice is to write a research project that will go one year beyond the maximum time allotted and ask for the maximum length of time. This will allow you to have enough time to complete most of the project, even if things don't go exactly as planned. Moreover, if the experimental plan is carefully thought out, then the preliminary data for the renewal will be ready to go.

Modular Budgets

NIH grants use modular budgets for grants that have direct costs of $250,000/year or less, in which you can request up to 10 modules of $25,000 each for each year. Now that you have come up with a number for your first-year budget, round up to the next-highest module and use that number for each of the years. A sample of the modular budget page that corresponds to the detailed form in **Appendix 3-3** is provided in **Appendix 3-4**. Specific information about using the modular budgets and samples of modular budgets can also be found on the NIH website.

THE REVIEW PROCESS

Earlier in this chapter it was suggested that the best way to learn about writing grants was to serve as a reviewer of grants. This section will provide you with a view of what happens during a study section or scientific review board meeting. The process has three stages: review; panel discussion and scoring; and funding

Review

The review process typically occurs over a five- to six-week period from the time that the reviewer receives the "box" of grants until the panel meets. Like you, reviewers have labs to run, committees to serve on, and other responsibilities. Reviewers will spend hours reading your application and writing the review. The reviewer's goal is to rank your grant among the group of grants that she is responsible for presenting to the committee. Significance of the problem at hand, quality of the preliminary data, feasibility of the experimental plan, and the perceived ability of the investigator to carry out the research plan are all points that are considered when the grant is reviewed. Sound familiar? In writing the review, she will attempt to highlight the strengths of your proposal and point out the weaknesses. The written critique is both for you to read and respond to if your grant does not get funded. The review also serves to refresh the reviewer's memory during the panel session so that she can describe and present your grant to the committee.

Panel Discussion and Scoring

Typically 20–25 scientists make up a review panel. Following opening comments by the panel chair, many agencies request that the review panel only discuss those grants that they feel are competitive for funding. For the NIH, this process is called streamlining, which used to be termed "triaging." The goal is to remove 50% of the grants from the discussion, allowing more time to prioritize the most competitive grants. In review panels where streamlining does not take place, grants that fall well below the competitive range are often only discussed for a brief period.

At the NIH, grants are scored on a 1.0–5.0 scale, with 1.0 being the best. The table below is the suggested "priority scoring" system used by the NIH. Remember, the reviewers' job is to prioritize or rank their applications with each other and the qualifiers shown in the table. At the onset of the discussion, reviewers will be asked to present a score (e.g., 1.7). The primary reviewer will present a description of the project and her critique. At the end of this presentation, which could be as long as 10 minutes, the chair will ask the secondary reviewer to state his view. Often the secondary reviewer will agree and say so. The secondary reviewer may highlight additional strengths if his enthusiasm is higher than the primary reviewer's, or may point out certain weaknesses if his enthusiasm is lower. A third reviewer will also be asked to comment. After the comments from the reviewers are finished, the floor is opened to the panel. Comparisons to other grants or discussions may also be made in an effort to rank the grants in an appropriate manner (conflicts of interests not withstanding). At some point the chair will summarize the comments, final recommended scores will be requested from the reviewers, and the members of the panel will vote with one reviewer or the other or split the difference. Developing a consensus is not required to complete the process. After scoring the grant, the budget will be discussed and an appraisal of whether or not the amount of money requested is justified. Most agencies' scientific review groups follow these procedures.

NIH SCORING SYSTEM

Score Range	Qualifier
1.0–1.5	Outstanding
1.5–2.0	Excellent
2.0–2.5	Very Good
2.5–3.0	Good
3.0–4.0	Satisfactory
4.0–5.0	Fair

Funding

All agencies send out some notification about how well or poorly your grant was rated. In the NIH system, you receive the averaged priority score (times 100) and a percentile rank. Percentile ranks reflect how well your grant ranked as reviewed from that study section only. Because priority scores tend to drift in one direction or another, the NIH determines percentile rankings from three meetings of that study section. Lower numbers are better for both. To be fair to all investigators, the NIH funds grants by percentile and not priority score. Percentile ranking prevents one study section from funding a larger percentage of its grants by just giving better priority scores. Thus, a priority score of 168 may fall into the 8th percentile in one study section and the 22nd in another.

Unless your percentile rank is outstanding (3rd percentile, for example), you will not know just by looking at the numbers if you will be funded. The NIH requests that you wait to see the summary statement (the review) before contacting your program officer. Your summary statement (sometimes referred to as a pink sheet because it used to be printed on pink paper) will come about 6–8 weeks after the panel meeting. If your percentile is low (a good number), read your summary statement and call your program director to find out about funding. If your score is in the non-fundable range (like 30th percentile or higher) or your grant was streamlined, then read your statement carefully. If you get mad, then do not call the program director until you have calmed down. Program directors are there to help get your work funded, so keep that in mind. Once you get your program director on the telephone, be prepared to discuss constructively how you can improve your grant application.

Even if your grant does very well, the NIH has a second round of review called "Council." Each NIH institute has its own Council, consisting of lay people, scientists, and administrators. Council has its biggest impact on grants that straddle the funding fence. Your program director will prioritize her grants that are in this range and seek to have as many as possible awarded funding by the Council of your institute. Each of the institutes has its own procedures for funding grants that are on the border. Your program director will tell you how to proceed. In rare cases, your grant may be targeted for funding at the end of the fiscal year (Sept. 30th), and you will get a phone call just before that date so that the paperwork can be put through. So, plan to be around a telephone to get the call from the agency if your grant score is close.

Resubmission

Sorry, but we've all been here. If your grant did not get funded, all is not lost. The NIH estimates that it takes about two submissions of a proposal to get a grant funded. You must therefore prepare for a resubmission of your proposal. To begin, go through your summary statement and identify the key points raised. When you discussed your grant review with your program director, she may have given you some clear suggestions about ways to strengthen your proposal. She may have been present during the review of your grant, so she has firsthand knowledge of the criticisms that were made during the discussion. If a low number of publications was the problem, you must deal with this issue prior to resubmitting your grant by getting your results published. If a lack of preliminary data is an issue, you must get the experiments to work, find a collaborator who can help you with the experiments, or both. It is not expected that you solve all of the issues, but you will need to solve or at least address the major ones and some minor ones, too. It is also OK to delete an Aim, especially if the reviewers hated it. If you still want to keep that Aim, you will need to be able to provide compelling new data to support its importance or feasibility. After you have considered the problems, take your summary statement to colleagues who have study section experience and ask them to help you. Yes, it may be embarrassing, but it is the only way to improve your score and move on.

> ▪ Remember, your colleagues are concerned about your getting funded, and they realize the road may be bumpy. They will help you if asked.

Revised NIH grants are allowed three pages for the investigator to respond to the previous critique. Here is your opportunity to state how you improved the grant by following the study section's advice. This is not the place to tell the study section what you really think and how insulted you are about your review. They don't care! NIH reviewers will have the previous summary statements, so they are fully aware of what the review said. Sometimes there are numerous points to discuss. If this is the case, you should break them down by Aim. Highlight only the major points. If one of their suggestions improves the grant or actually worked and there is preliminary data to support it, say so. If you don't like one of their suggestions but the reviewers state that they think it is important, consider ways to incorporate it in the proposal. You may actually find that it really was a good suggestion. If a suggested experiment cannot be performed, explain why and then provide an alternative. Sum up the changes in the end. Be positive in your responses to the suggestions, because this is probably the first thing that the reviewers will read.

Our *Next Wave* article, *I Can't Believe They Didn't Like It — Grants*, discusses responding to grant reviews and provides additional detail and discussion to this section. The article is presented in *The Extras Section*.

OTHER TOPICS ABOUT GRANTS

Human Subjects

The use of human subjects in biomedical research receives careful scrutiny by both the host institution and the funding agencies. The procedures, which have been developed over the past five decades to address egregious practices, are important for the protection of the rights and health of human subjects. If your work involves human subjects it is incumbent upon you to follow *all* of the rules and guidelines of your Institutional Review Board. The U.S. government requires that all key personnel on projects involving human subjects complete courses in protecting human study volunteers and pass one or more exams prior to the beginning of such studies. The book, *Protecting Study Volunteers in Research: A Manual for Investigative Sites*, by C. McGuire-Dunn and G. Chadwick, provides descriptions of the laws and the processes that one must follow. The NIH has recently enhanced its review criteria for studies involving human subjects. Each institution has an Institutional Review Board (IRB) and a Human Investigations Committee (HIC). These committees review protocols and procedures and advise the investigator on how to comply with federal and local regulations. If you use human subjects in your work, you should learn all your local rules, follow them, and, most importantly, get the paperwork completed long before your grant is submitted.

Vertebrate Animals

Many biomedical researchers use vertebrate animals as a model for human disease or injury. Such studies are under substantial pressure from the community to be "necessary" and subject the animal to as little pain and suffering as possible. The benefit of animal-based research is without question. To be responsible, a scientist using animals must take the community's concerns seriously and follow the recommended practices and guidelines established both by governmental sources and the local Institutional Animal

Care and Use Committee (IACUC). Your experimental goals, protocols, and the number of animals that you will use are reviewed by your local IACUC. Grants will not be funded without such approval. You and your laboratory personnel will also have to complete specific courses and pass exams to be certified in the use of research animals. If you are unsure how to write a specific aspect of these proposals, ask one of your senior colleagues or the local veterinary doctor how to write the section. Again, prepare early.

When to Put in Your Second Grant

Your start-up funds may be sufficient to allow you to coordinate two projects. If this is the case, and you have enough help to allow you to do this successfully, then you may be able to put in grants on separate subjects. While this idea is tempting, each will require the same level of data and preparation to be successful. If you choose this route, do not sacrifice your effort on one grant so that you can put in two. It will be best to get one funded, begin your independent career, and worry about a second grant in a year or two.

In deciding when to expand your program and write the second grant, you should consider what progress you have already made. One criterion that the study section will look at is whether or not you have been successful/productive on your first project; i.e., did you publish some papers, is the work in good journals, etc. If you have not been productive, it will be difficult for the reviewers to justify giving you additional resources no matter how good the project reads. If you have been productive, this too will be noticed and you will stand a better chance of earning a high priority score (a low number) on the new grant.

CHAPTER 4

MANAGING YOUR LABORATORY

"Nice lab. Your lab must be bigger than mine, 'cause I didn't have room for the TriPlex GammaZoid," they will say. They may also ask: "Have you hired anyone yet? What about that postdoc? What about the new student with the curly red hair?" You may wonder these things as well. Additionally, you may wonder, "Do I have enough money for all these people? How will I divide up the project? Where will I put them all?" Oy! Welcome to running your laboratory.

Lab management is a complex task, akin to running a small business. Your students will even give your lab a name when they answer the phone: "Hello, Bosslab." To run this business, you will need not only your skills as a scientist but also some business skills. Such skills include: budgeting and spending money; hiring and firing staff; supervising staff; establishing lab policies; adhering to institutional guidelines; and providing support and guidance to your staff. This chapter will focus on some of these issues and attempt to provide you with an approach to gaining these skills. The specific topics covered here will be:

- Money management
- Hiring and firing staff
- Supervising staff
- Establishing laboratory policy

MONEY MANAGEMENT

"Money Management," you say. "Hah! We just got $150,000, so let's not waste any time. Let's purchase pre-made solutions and kits for every experiment." Unfortunately, if this approach to spending is used throughout the entire year, you may end up saying, "We can't buy anything until next month because we are $20,000 in the hole." When your lab grumbles at your budget restrictions, you realize that managing money is perhaps the worst aspect of running a lab.

You may think, "We should all be given enough money to do the research that we can think of. After all, we're biomedical scientists, and we're going to cure disease and improve the quality of life!" Unfortunately, Research Utopia University (RUU) doesn't really exist and money is the commodity that will allow you to investigate to your heart's content. From a business point of view, funds fall into two categories: renewable and nonrenewable. In the beginning, you will have start-up funds and, hopefully, you negotiated a solid package (see *Chapter 1*). Ideally, start-up funds will allow you to work for up to two years before you must rely on extramural funding for your salary support and that of your lab personnel. You should consider start-up funds nonrenewable, and spending them should be done with considerable thought and planning. In contrast, extramural funds typically provide a fresh pot of money once a year and are therefore "renewable" funds (at least until the grant runs out). While spending renewable funds should be carefully planned, there is more room to recover from errors in spending. Thus, while some of the issues in money management are the same for renewable and nonrenewable funds, there are clear differences and each will be addressed separately.

Budgeting Start-up Funds

There are three philosophical approaches to spending your start-up funds. Each depends on the time frame in which you obtained or expect to obtain extramural support. Approach 1 is the conservative approach. Here you plan a budget that stretches the full two years. This ensures that you can keep on generating data during the entire period of time and increase your chances of getting your first big grant. The liberal approach, approach 2, is to hope for the best. In this scenario, you are banking on the hypothesis that you will definitely get funded during year 1, and therefore you will liberally use your funds to rapidly accelerate your lab's research into the next issue of *Science*. The liberal approach also has the underlying theory that if it does not work out as predicted, someone — like your chair — will save you. The third approach is considered when you already have extramural funds in hand or expect to have them arrive soon after you start. In this case, you can use your funds to purchase a fancier piece of equipment, expand the size of your program, or attempt to branch out into a second area that will ultimately lead to additional funding opportunities. You may even be able to "bank" some of your funds for a rainy day or the chance to hire a new postdoc if the opportunity arises.

Regardless of your philosophy, you will need to consider how your money will be spent and how to budget it for the length of time that you wish to have it. It is best to break expenses down into one-time costs (equipment) and recurring costs (personnel and supplies). You should plan your recurring costs on a monthly basis. This way it is easy to add in funds that arrive at different times during the year. There are a number of software programs, such as GrantsManager and Personnel Manager (www.northernlightssoftware.

com) that you could use. Microsoft Excel spreadsheets are also very useful but require a little more skill to set up the monthly incoming funds and outflow. If you are not a computer type of person, green ledger paper and pencil may be the best way to map it out. Just don't wait until your start-up fund is nearly used up to start budgeting!

Begin by taking your total $$$$ (direct costs) amount and subtracting the costs for equipment. If you plan on "banking" some of the money as suggested above, remove this sum from the balance as well. Divide the remaining funds by the number of months you wish your start-up funds to support. This is your monthly budget. Doesn't look as big as it did before, does it? Now you should allocate these funds into the main categories for running your lab. Primary categories of expenditures are established in the non-modular NIH budget sheets (NIH 398-Form page 4 as shown in **Appendix 3-3**), and include:

- **personnel expenditures** (salaries and fringe benefits);
- **equipment** (typically items >$1,000);
- **supplies** (everything that is consumed by your lab);
- **other expenses**
    - **animal costs** — this is a tricky category and includes not only the cost of the animals but also the costs of maintaining the animals, referred to as per-diem cage charges;
    - **service contracts** (surprisingly expensive — you will be stunned);
    - **fee-for-use services** (flow cytometry, mass spectrometry, microscopy)
    - **page charges** (for that *Science* paper)
- **travel** to meetings

How much does it cost? Information on personnel salaries will reside with your departmental business manager. He can also provide you with the companies that provide a standard discount to your institution. This person can also help you with your budget. Department business managers have seen it all. They have listened to all the tales and woes of poorly executed financial plans and have marveled at those in which the faculty member has gotten something for nothing. It is also in their interest to make sure that you do not overspend. They are therefore likely to give you the best advice and follow up on information that you need to make your decisions. Your departmental business manager is your friend.

Budget Category Details

Personnel

Salaries will always be your largest recurring cost and should be considered first. In addition to the wages that you provide, most positions come with fringe benefits. Fringe benefits include Social Security benefits, health insurance, retirement programs, courtesy scholarships, etc., are included in compensation packages for employees. Fringe benefit rates may vary depending on the type of employee being hired (e.g., Lab Technicians vs. Postdocs). Because students aren't eligible for most benefits, their fringe benefit rates may be 0%. Graduate student stipend amounts and their tuition costs as well as who pays for these expenses vary among institutions. Check with your business manager to find out when you would have to begin to bear the costs of having that new graduate student with the curly red hair work for you. It also may be possible to defray your costs on this student or a postdoc if your department or program has an NIH training grant. Your chair or the "Program Director" (Principal Investigator) of the training grant will be able to provide you the most accurate information about the availability of such sources of funds.

After budgeting the personnel category, you will find out how much money you have for extraneous equipment or for extra bells and whistles on the important equipment that you need to buy.

Equipment

Your major equipment purchases were included in the initial start-up budget that you negotiated. If you saved some money on your equipment purchases because you were able to cut a deal with Joey Spinner, the centrifuge rep, you may want to reserve some funds for the year in case your talented new student with the curly red hair breaks the loader on the TriPlex — twice. You may also need some new equipment as the assays in your lab change or the number of people you hire increases. So, budget a few thousand per year for repairs and new small equipment.

Supplies

This is the second largest recurring cost. Find out how much others in your field spend on supplies per person per month in your department. A general figure for a molecular lab would be between $600 and $1,500/person/month. However, if you require specialty reagents, such items can greatly affect your costs, so be sure to add these into your budget. The costs to purchase animals should be included here as well. You will find the cost of using animals is extraordinarily expensive, and the monthly per-diem costs are always a surprise (see Other Costs below). Establish a budget using these figures, and don't forget to include yourself as a supply consumer.

The surprise in this category is the hidden shipping charges. Each item that you order comes with substantial shipping charges. If you place 20 separate orders for reagents that must be shipped on dry ice over the course of the year, this could cost you a thousand dollars. However, if you can combine a number of items into the same shipment, you may be able to save enough money so that you get the fifth item for free. See the section below for additional ways to save on supplies.

Other Costs

Other costs vary greatly depending on your use of animals, research core facilities, service contracts, or user fees on shared equipment, etc. Finding out the costs on these items will help in making your money go a long way. For example, it may turn out that you pay a portion of a service contract for a particular instrument depending on your percentage usage. If there are multiple instruments and these charges are calculated on each instrument separately, then using one of the instruments exclusively may end up saving you money at the end of the year.

Animal housing costs can quickly eat up any budget. Cage costs are calculated on a routine schedule. Find out when the cage census is taken in your facility and plan your experiments, if possible, to most effectively conserve cage costs. Additionally, you should manage the size of your breeding colonies so that you do not end up with too many unnecessary cages. It may seem like a little bit of money each time, but at the end of the year it could add up to thousands.

If you plan on using the local core facilities be sure to find out ahead of time what the typical hourly or procedural charges are for the experiments you want to do. The overall costs of using core facilities may ultimately have a large impact on your budget, so proper planning will be important.

Travel

From the ski lifts of Keystone Resorts Conferences to the shores of the Cayman Islands Convention Center, scientific meetings serve as venues to exchange results and ideas. Meetings function to bring everyone up to date, and allow you to meet and interact with your friendly (and not so friendly) competitors. If you are fortunate enough to be invited to speak, these forums provide you with the best advertising for your research that you can hope for. Thus, even a beginning Academic Scientist should go to meetings. The advice here is to go to the meetings that will provide you with the best interactions and information you need to include in your grants and papers, even if the meeting is in a dull place. Of course if the meeting is in an exciting place ... enjoy.

Renewable Resources

Five years of RO1 funding! Yippee, new computer here we come! This may be the first thing that comes to mind after you see your superduper grant priority score and percentile rank or get a telephone call from someone at the granting agency giving you good news. As you sober up, the second thing that may come to mind is "We only have 4 years and 3 months to produce enough data and publish papers so that the grant will get renewed!" Unfortunately, this thought does rush in and spoils some of the fun in getting funded. Hence, good planning in Year 1 will allow you to move steadily toward that renewal without fears of spending Year 3 money in Years 1 and 2 and finding yourself short of funds in Years 4 and 5. This scenario happens. Often.

When you submitted your grant application, you made a budget, and, like every budget in the research universe, it was somewhat "padded." NIH modular budgets allow for invisible padding in all years, so that when they cut your budget by a module or two, it shouldn't hurt as much. Budget cuts obviously hurt grants that have less funding. It is important to realize that when a grant is cut by 20%, you don't actually cut your own salary or that of your staff by that amount. The net effect is on the total number of personnel, supplies, etc., — all categories that allow you to do the work.

Budgeting and budget outlines are necessary to organize your actual spending. Once again, to start budgeting, remove the equipment amount from the total yearly budget and plan your monthly expenditures. Start with personnel, followed by supplies, other costs, and then travel. You should plan each budget year separately because the funds will renew. In many cases (NIH grants for example), monies can be carried forward to the next fiscal year. If this is the case, then thriftiness early in a grant may allow you to accelerate your progress by adding people to the project in the second year. While this may be desirable, NIH makes it difficult to carry forward more than 25% of the grant funds from one budget year to the next. So, plan carefully.

In case you overspend your funds in one year, you should assume that your institution will reduce your renewal funds for the next year by that amount. You may even find that if you overspend extramural grant funds at the end of the funding period your institution will bill you for both the direct and indirect costs associated with your spending (see *Chapter 1* for definitions of indirect costs). Therefore, it is in your financial best interest to monitor your expenditures, especially as a grant is expiring.

If you have multiple sources of money, then viewing the whole amount on a single spreadsheet may be useful. You need to remember that grant funds are often restricted to certain projects, and therefore spending monies from one to help another is not appropriate.

Monitoring Costs

Your institution will most likely provide you with some form of expenditure report. Unfortunately, these reports are almost incomprehensible for scientists. Sometimes, departments provide their faculty with financial statements that include encumbered costs — funds that you have committed to spend but haven't yet. Even in the best case, these statements will be behind by at least a month or two. So, seeing that you have $30,000 left on a grant account that ends next week, and spending it that minute may not be a good idea until you check with someone in the department who is responsible for the bookkeeping or accounting of these funds.

Some people like to keep their own records of supply expenditures. This is a good idea but needs to be done in "real time" or you will quickly fall behind. This is an ideal job for a work-study student. This could also be a great time to sink if you end up doing it yourself. So be wary of the amount of time that you devote to this endeavor.

The key to using your funds to your best possible advantage is to establish a spending plan within a budget that is worked out with input from faculty and your business manager. While the best scientists in your institution may overspend their budgets every year, they also may have to pay the consequences occasionally by having to terminate the employment of productive workers. So, budget wisely. As noted earlier, grant accounting software is available, as are spreadsheet programs.

> ▪ Don't get overwhelmed or sidetracked with the tiniest of details in managing your budget. Establish a budget plan and follow it as closely as possible.

> ▪ If budgeting is not your thing, seek help in establishing your budget from your departmental business manager and try to follow his advice.

HIRING STAFF

Who should my first employees be? Research technicians, postdoctoral fellows, senior research associates, and graduate students make up the basic categories of lab staff. Each category comes with its advantages and disadvantages. Here is another opportunity to find out from other investigators in your department how they staff their labs, remembering that, depending on the success or seniority of the investigators, their approaches to hiring will be different. Senior investigators are likely to employ all three categories, while junior faculty may have only had time to recruit a technician and some students. A key point is that the more people you have in your lab, the less time you will have for your own experiments. You are your own best asset and need to be in the lab as much as possible. As a new independent investigator you need to remember that putting your career and reputation fully into the hands of others does not make sense. So, choose wisely.

Postdoctoral Fellows

Postdoctoral fellows make up the strongest group of possible employees. They have been trained in numerous technologies and in many cases want the training that your lab will provide so that they can get your job or one similar to it. Because beginning Assistant Professors have not established a track record in training postdoctoral fellows and helping them find jobs, they have trouble in recruiting talented new PhD graduates. Sometimes a new investigator can get lucky and find that the perfect postdoc is knocking at his or her door. You should be wary of taking a postdoc who is not well trained and was not productive as a graduate student. Problems that a postdoctoral candidate for your lab had as graduate student don't go away because he graduated. This means that you should review the applicant's CV carefully. It is important to get letters of recommendation from the candidate's advisor. You may consider calling the advisor and talking with him/her directly. If you are unsure about a future prospect, ask a senior colleague to read the letter in case he sees something that you do not. It is also important to remember that hiring postdocs is also a commitment to helping in their careers (see *Chapter 8, Mentoring*). If you do hire a postdoc, be sure to discuss the project goals ahead of time and provide an honest view of how you anticipate funding the position and for how long.

Finally, if you hire someone who has yet to successfully defend his dissertation, you may have to deal with his writing the final drafts of the thesis and manuscripts for publication on your time. This can be a difficult situation because you need him to be productive in your environment. It is best to wait for him to defend his thesis before paying him to work for you.

Graduate Students

Graduate students make up a major labor force in many laboratories. Like postdoctoral fellows, graduate students are being trained to carry out research. The symbiosis between the student's "pair of hands" and the "training environment" that you provide is critical both to his success and to yours. Because graduate education programs last between 5 and 6 years, graduate students provide the greatest continuity to a research program.

Beginning graduate students vary greatly in their level of expertise and experience. Some are as green as grass while others have been well trained as research technicians. Which ones should you take into your lab? This is a never-ending question. Most of the time, investigators do not have a choice over which student to choose, as the students typically do the choosing. The investigator gets to say only, "Yes you may come," or "No, I think that it will not work out" or "My lab is too crowded."

> **Survey Says:** An Assistant Professor — "The ability, or lack thereof, to attract graduate students in the lab has been the single most important factor in establishing a productive lab ..."

How do you decide whether the new student with the curly red hair is the one? There are two factors: You and Them. To decide, you must first take into consideration the number of people in your lab and how much time you can and want to dedicate to training *this* student. Students require an enormous investment of your time. The more students that you have in your group, the less time you will have to carry out your own experiments and to write your grants and papers. Therefore, the number of students is a key issue. Realistically, a new faculty member should have no more than two students for the first two or three years of his/her career. This number can change as the number of *funded* projects increases and the number of *published* papers increases.

The second factor is that you need to determine the prospective student's level of seriousness. Is the student serious about science? Does he read what you assign? Did he blow up the GammaZoid or destroy other expensive equipment in the lab, twice? Can he perform simple experiments and get them to work? Are the data that he generated "figure quality"? The answers to these questions are easily ascertained by evaluating the prospective student's performance during his laboratory rotation. If he is not serious, does not read what you assign, breaks equipment, and can't run a simple experiment, DO NOT under any circumstances take him into your lab. You will regret it. If, however, he fulfills these simple criteria, go after him; he will be a great colleague to work with while he earns his degree.

Research Technicians/Specialists

Research technicians can provide continuity and great service to your lab. In many areas of the country, it has become increasingly difficult to find and retain research technicians because of the competition that exists from industry and because of the range of career options that are available to college educated life science majors. Why should you hire a research technician? To carry out your experiments? To prepare reagents for the lab? To clean the glassware? To order supplies? Run common use instrumentation? Before you hire a research technician, answer these questions. If you want research technicians to collect data, then you may want to hire individuals who are more experienced than those you hire to prepare general lab reagents, wash the glassware, and order supplies.

> **Survey Says**: A tenured Associate Professor — "... the key to success is getting productive people into the lab early on." An experienced senior technician is invaluable.

"Some labs function quite well by hiring very bright, new college graduates who want to work for a year or two while determining if they want to apply to medical school or graduate school. The downside to this approach to staffing your lab is that the constant turnover can be time-consuming as far as hiring and training (plus all the going-away parties) is concerned. However, these technicians can be excellent employees because they are energetic, intelligent, and have professional career objectives that lead them to make the most of their time in your lab.

> **Survey Says**: A tenured Professor — "From the very outset, train an intelligent lab manager who will stay with you for some years, manage all of the time-consuming nitty gritty of the lab, and will train and watch over students and postdocs in ordinary lab techniques."

On the other side of this coin are the "professional" research technicians. These scientists love research and doing experiments. They often are uninterested in writing papers or grant applications and therefore did not pursue PhD degrees. They also provide long-term continuity and are likely to be the ones who can collect data for your grant proposals and manuscripts. They may even be able to help train your students and fellows in the technologies that you do. Occasionally, you may be able to find a senior technician who will bring new technology to your lab. If the one you hire is experienced in your field, you will save substantial time in the initial training period. Even though the salary cost is higher, a senior researcher can instantly boost the productivity of your lab, work through new procedures, and launch your career.

> ▪ While you are determining the best personnel to hire for your lab, keep in mind that you will need to keep a bench for yourself and work there consistently. Not only will you be assured that some of your lab work is going forward, but you will have the opportunity to establish work patterns and lab procedures by example to the other members of your lab (assuming that you have work patterns that are worth emulating).

If you hire someone who is currently working at your institution, you may find that your new hire has 6 weeks of vacation that he has not used and he wishes to take all six soon after joining your lab. Find out how much vacation time he has before you hire him and talk with your department business manager about who will pay for the vacation time. You may even consider asking him to take some of his vacation days before joining your lab so that you will not be paying for vacation days that were built up somewhere else.

How to Choose

Hiring Research Technicians

Educational background, employment history, and research experience of potential candidates are the initial traits that you will be able to evaluate from employment application forms. Obviously, you want the person with the best education, steady employment record, and the most research experience. The job interview is where you will be able to get real information about his or her experience and whether or not your personalities are compatible. From your list of candidates, choose a small number of people to interview. It typically takes about 30–40 minutes to interview someone, so plan your time accordingly. Your interviews should be structured and the same for each person. **Appendix 4-1** provides a worksheet with important issues and questions for each candidate. You need to evaluate each candidate's qualifications, and his or her compatibility with your work-management style.

The applicant's research experience is the easiest aspect to evaluate. Ask him to describe what he has done. To get a feel for how well he participated intellectually in the project, ask him to provide the scientific rationale for the project and the experiments that he has performed. Ask him to list the techniques that he likes to do and those that he does not. This is important because he may be leaving his current job because he is bored with a particular set of assays. This question will also help shed light on the techniques that he may have watched vs. the ones that he actually performed.

> ▪ Communication is the key to any good relationship.

You must be able to understand the candidate and, most importantly, the candidate must be able to understand you. The quality of communication can be judged by paying attention to the dynamics of the conversation during the interview. Is the candidate always talking? Are you? Does the candidate answer the questions that you ask? To test how well the candidate understands your research, provide him with a short description of your project(s). Keep it simple, highlight the rationale for the project, ask him basic questions along the way, and try to engage him in the discussion. If his eyes glaze over, he is not for you. This is also your opportunity to recruit him to your program if you think that he is the ONE. If you work with animals, be sure to ask if he is OK with animal research. If you use radioactivity, hazardous chemicals, or ebola virus, now is the time to inform him of the lab environment. Because all job candidates are very concerned about salary structure, you should know the range that the position offers. You do not have to provide prospective employees a salary number during the interview. If you really want him, you may ask what he is earning now or what he expects to earn, as this will give you an idea of what you will need to spend to attract him to your group.

If you have specific ideas about work hours and lab participation, you should explain this to the candidate during the interview. He may ask you about authorship issues, lab duties, and participation in science activities of the department. Be ready to answer clearly and candidly. You also may want to give him a lab tour. This will allow you to see how well he knows certain types of equipment, and it provides him with a view of his possible future home.

Reference letters and phone calls to previous employers or mentors are critical to making your decision. DO NOT under any circumstances hire someone without checking into the candidate's previous employment or research experience. Yes, we said DO NOT!!! You will be surprised at the number of people who do not call the previous employer. Failing to do so is a disaster waiting to happen.

The more information you can get, the better the chance that you will not be making a mistake. Ask references about the technology that the applicant performed and the quality of the data generated. Also ask: did he got along with labmates; did he follow instructions; was he independent; did he break equipment; did he keep a good lab notebook? Lastly, ask "Would you hire him again?" Hesitation to answer this question may be the key that you need to decide between person A and person B. Hesitation on any of these points may provide you with clues to behavior traits that you may think are important.

After completing all your interviews, you will be ready to make a decision. Most institutions have a human resources department that has guidelines that must be followed to ensure that equal opportunity policies were followed. Check with your business manager to find out what these policies are and how to follow them. The human resources department may also have protocols that should be followed when offering someone a position in your lab. When making an offer, be sure to clearly state the terms. Your business manager or human resources office may have a form letter that you can use. You may want to indicate a probationary period in the letter. Find out the acceptable time frames for this period from your business manager. A sample form letter is included in **Appendix 4-2**.

Hiring Postdoctoral Fellows

There are several key issues to consider when hiring a postdoctoral fellow: competence, productivity, motivation, and compatibility. Was the person productive as a graduate student? If she had two or more publications in quality journals in her field, the answer is Yes. If she had one publication, was it extensive and in a high-quality journal? If so, then the answer is still yes. If she had no publications, this is a problem and could indicate a poor level of productivity and/or competence. If there is a paper in preparation, will she be finishing this work on your time?

To answer these questions, invite the prospective postdoc to your lab for a visit. Have the candidate present a short seminar on her research. In addition to your lab members, invite other faculty members or postdocs from your department to the seminar. Additional scientists in the room will provide a more robust atmosphere that may be helpful in recruiting this person. Reference letters from her advisor and other evaluators are very important. A phone call to the previous mentor should be made even if you have a glowing letter. By using the candidate's CV, seminar, and phone calls to her advisor, you should be able to judge all the areas discussed.

Recruitment of a postdoc may be difficult, but if you have a chance at getting a good one, you may want to ask your chair for supplemental funds to help recruit the person to your lab. While you may not be successful at getting the money, it's worth a shot. To help you recruit postdocs, your school may have an "Office of Postdoctoral Education" that will have lots of information about recruiting postdocs to your institution. They also may have information about living in the area, statistics on what the postdocs do after they leave your institution, and where they all are. To help with recruitment, it would be a good idea for your candidate to meet other postdocs in the department.

Our *Next Wave* article **Staffing Your Lab** provides additional points and discussion on this subject and is reproduced in *The Extras Section*.

MANAGING YOUR GROUP

Supervising Staff

"All I have to do is to outline the experiment on paper and my graduate student will just do it like I did when I was a student, right?" Yep, it will help if you pray a lot too. Your advisor probably did a little more to get you to perform well than you remember. While each person has different capabilities, experiences, and talents, you must decide how much supervision each person needs in order to perform up to your expectations. Complaining to other faculty members that a student or technician is not generating enough data or that his or her preparations are not clean enough will not improve your staff's productivity. Remember, productivity is the name of the game. There are several approaches to supervising. Whole lab meetings, smaller working group meetings, and one-on-one meetings allow you to assess the progress of your research staff. Each of these formats has its benefits, but all take your time so scheduling a week full of meetings may not be in your best interest if you are performing your own experiments. Below is a discussion of each format.

Lab Meetings

Lab meetings have different formats depending on the size of the lab. In small labs, everyone can present what they did for the week and place it in the context of their projects. If your lab is young in experience, it is also a good idea for you to discuss a paper in your field as well. This approach ensures that at the end of the year everyone has been exposed to the major papers that you think are important. An option other than doing both during the same meeting is to have people alternate between presenting a paper and their own work. This will also teach your students how to critically evaluate a journal article. As your lab increases in size, the number of people presenting will have to be reduced so that you do not use up most of the day. You also should reserve some time in the lab meeting for people to talk about lab issues. This way the complaints can come out in the open and serve as a little group therapy, and problems can be resolved quickly.

One-On-One Meetings

Individual sessions with staff members provide the highest level of supervision, allowing you to discuss experiments and goals with your staff. These should be arranged on a regular schedule and you should keep to this schedule. You should make your staff keep to the schedule as well. It is best to have everyone prepare a list of what they view as their weekly or biweekly goals for you. Weekly records of goals is the best way to monitor productivity and for you to judge whether or not the two of you are communicating.

Small Group Meetings

An excellent way to direct and supervise research efforts as your lab gets larger or as the number of distinct projects becomes greater than 1 is to meet with a subset of your lab members. In this context, 2–4 people who are working on a project or who have overlapping goals meet and discuss their progress in a similar manner to the one-on-one meetings discussed above. The advantage of this type of meeting is that everyone working on the project is made fully aware of everyone else's progress and the status of reagents and techniques. Unlike the larger lab meeting, small group sessions allow for greater interactions and exchange of ideas among your lab members.

> ▪ Ultimately, you will find the style of lab meetings and supervision that most suits your personality. The goal of course is to allow your students, fellows, and independent technicians to explore their own ideas and to contribute their brilliance to your research program.

Problems

When you find yourself complaining to your colleague, Dr. Pat I. Ence, that your students don't seem to be doing what you want them to do, you know you have problems. Some common problems with solutions are discussed below. Dr. Ence may have some advice too.

Not Listening

What do you do if someone is not following your directions and advice? There are only three reasons why someone is not following your direction. The first is that the person does not understand. If this is the case, you may need to write out a step-by-step protocol for him to follow. Clearly written instructions avoid the "**read-my-mind**" syndrome that faculty are said to have when rapidly describing experiments. You may even have to watch him do the experiment to see what he is doing wrong. As always, if you know how to perform a complex procedure, it is best to demonstrate how to do it, even if it means demonstrating it two or three times.

Knows Better

Another reason why he may not be following your directions could be because he thinks he knows better. If he gets the experiment to work and the controls are satisfactory, then you may have to adjust your thinking. However, if the experiment doesn't work, you will have to explain in *very clear* terms that you expect your experimental design to be followed and only changed if it does not work or at the very least after consultation with you.

Do It Right

Your lab motto should be "**Just do it right ... all the time.**" The failure to do it right the first time or every time can be due to many things. Sometimes people simply try to cut corners. Cutting corners wastes everyone's time and your money. Other times, your student may not really be set up to do the experiment and is scrambling to get the reagents together in time to do the experiment. When he finally has prepared or collected all the reagents, the time point is past or the sample is now worthless. In the worst-case scenario, the student or tech really didn't pay much attention when he was shown the procedure, and therefore needs retraining. A session with you watching him perform the experiment is one way to get him to do it right.

> ■ In all cases, you need to be direct, tactful, and honest with the person who is not following your advice. Do not become unprofessional and complain to one employee in the lab about another. This will not go over well, and they may wonder what you are saying about them to the others.

Not Putting in the Time

What do I do if staff members are not putting in the time necessary for completion of the project? The answer to this common problem depends on the staff member's level of employment and how he is spending his time. Research technicians are regular employees and should be working a 40-hour week. While many technicians work more than they are paid for, this should not be expected. You should plan the experiments with them so that they can be completed within the normal workweek. If they are not working 40 hours, then it is up to you to document this in writing and tell them to do so or risk losing their jobs.

Postdoctoral fellows are expected to be professionals and to complete their projects in a reasonable time frame. This means that they should be working 40+ hours a week. If they are not productive and not working 40 hours a week, you should consider letting them go. Even if the postdoc has his own money, a nonproductive, or in this case, a lazy postdoc will use up your supplies and importantly your time. In the worst case, such an individual may lower the expectations of others in the lab about what needs to be done and what level of performance is expected. After all, what will you say when asked to provide him with a letter of recommendation? You certainly should *not* say you would hire him again!

Graduate students fall into the most complex category. Classes, teaching requirements, and preparation for qualifying exams take up much of their time during the first two years of their scientific lives. The general rule is that they should be working in the lab when they are not in class or have other academic responsibilities. If you have a student who has completed her first two years of course responsibility and she is not working 40 hours a week, this is a problem. Because graduate students are (supposed to be) monitored on a regular basis by thesis committees, you may be able to use the committee in a "passive aggressive" way to change her behavior. However, most people respond to direct and honest discussions. The questions to ask a lazy student are: Do you really want to do this as your profession? Do you want to spend more years here than are necessary? While these questions aren't necessarily wake-up calls, they will certainly get the student's attention. If the direct approach does not work, then talk to the graduate program director. She may have some advice on how to proceed. Because of the time involved in completing a PhD degree, there is considerable inertia in forcing a student to switch labs. However, the earlier a problem is identified, the better for both you and the student.

Wasting Time

There is also the possibility that your staff is working 40+ hours / week but they are spending their working time on the Internet, writing emails, organizing fantasy football games, or just talking. This too is a problem that will reduce their productivity and ultimately yours. How do you deal with these issues? Specific issues and potential solutions are presented below (see Policies). It is always important to remember that having a relaxed atmosphere is important for the overall happiness of your employees. Thus, finding the balance between productivity and fun in the lab is of utmost importance.

If problems persist, it is important to use tact in your approach to correct them, as you may not want your staff to quit. It is considered unprofessional to chew one employee out in front of others. So, take the person to the woodshed (your office) and firmly indicate where he went wrong. If the problems persist, document the problems in a brief letter to the individual.

Evaluation of Your Staff

Written and formal evaluations of your staff are important to them and could be very important to you. From your new friend, the business manager of the department, obtain the policies for evaluating staff members. This will probably be an established evaluation form for your use in conducting periodic, written evaluations. It is very important to follow the institutional and departmental procedures in evaluating your employees. Follow the policies and procedures for evaluation and employee feedback and you are less likely to do something wrong that will result in a time-consuming employee relations debacle. You do not have the time to do anything wrong, and now that you are the "Boss" you have many more opportunities to fall into deep holes that will take extraordinary amounts of time to climb out of. In summary, follow all of the institutional procedures regarding personnel, and document what you do.

Firing Staff

In the unfortunate event that you chose poorly and the person is not able to perform the work of your lab in an accurate or timely manner, talk to the departmental business manager (he is getting used to you by now) about the "due-process" procedures for dealing with the problem. The best result would be that after you point out to the employee exactly what it is that needs to be improved in his work, and give him a reasonable period of time to rectify the problem, the employee turns out to be an asset to the lab. However, if the due-process procedures are followed and the employee cannot improve to a reasonable level, then you have to have the support of the university to proceed in ending this person's employment. You should not proceed on your own in this process without the advice and assistance of your business manager or the human resources professionals at your institution. Written evaluations provide the necessary paper trail that you need to get out of this problem. Ask your business manager for copies of the employee evaluation form recommended by the Human Resources Department and use it.

Of course, you could just fire their lazy !#$#!*%@. However, if you do, you must be prepared to deal with the consequences of your institutional bureaucracy. One consequence may be that you find that your behind is dragged in front of a review panel for the method by which you dismissed your employee. You may have to justify why you fired this person as opposed to another in your group. Human resources policies are there to protect you and your institution from the legal consequences of hiring/firing decisions.

> ■ If you need to fire someone, **do it by the book**, and do it as soon as you think it needs to be done.

OTHER MANAGEMENT ACTIVITIES

Aside from scientific experiments, there are numerous tasks that must be performed for a lab to run smoothly and successfully, as well as to remain in compliance with local policies and guidelines. These tasks can be broken down into several groups: Institutional Procedures; Laboratory Organization; Spending Money; and Laboratory Policies.

Institutional Procedures

All institutions have procedures and guidelines for research involving human subjects, animals, hazardous biological organisms, hazardous chemicals, radioactivity, and restricted drugs. To use any of these, you must comply with federal, state, local, **and** institutional rules that are set up to protect you, your employees, and others from physical harm, and to protect you and the institution from potential lawsuits. If you use hazardous reagents, find out from the business manager where the offices are that oversee these policies. Get their current forms. It seems that every couple of years all the forms change and then you have to do it all over again when it is time for your renewal to use restricted reagents. For the most part, they will ask you to identify the reagents that you will use, how you will use the reagent, how you will dispose of the reagent, and what precautions you will take to ensure that neither you, your lab, or the janitor will be exposed to it. For protocols involving radioactivity, involving animals and human subjects, you may even have to take a short course or a short exam. Yikes! If you use human subjects or vertebrate animals in your research, be sure to allow plenty of time for revision of your protocols and approval (see *Chapter 3*). Talk with your colleagues to find out how they filled their forms in. Remember, save your innovation and creativity for your grant applications.

Spending Money

Everyone spends money differently. It is important to remember that start-up funds and initial grant funds give you the chance to establish your research career. Be conservative. Splurging on a fancy kit that gives no clear advantage except that you did not have to make up a Tris solution is not a good use of money. However, if you have to warrantee that the solution had certain consistent properties or was free of contamination, then this may be the only way to go. Other decisions involve bulk, smaller size, and pre-made reagents. As in the rest of the real world, buying science supplies in bulk is usually cost-effective, but there are a few principles to consider. The first is the shelf life of the item. If you will use the supersized, el bulko size of the item up before the expiration date, then the large size may be a good choice. The second factor to consider is whether or not you will in fact **ever** use it up. Having four 4-liter bottles of xylene sitting on the floor for 8 years is not a good investment (this advice is based on personal experience). The third is storage. While buying a 50-kg tub of agar may reduce the cost of bacterial plates to mere pennies, you may be faced with considerable storage problems in your 800 sq. ft. lab. Thus, consider the space and the footprint that storing bulk items may require.

Pre-made Solutions vs. You-make-'em Solutions

There are two arguments for buying pre-made solutions: (1) the solution is ready to go and guaranteed to work (by someone you don't know); and (2) if the solution is toxic in crude state, then purchasing this item in a refined state will decrease exposure of your lab personnel to the toxin.

Two easy examples where it may be better to buy the ready-to-go item: Acrylamide and Phenol. Both are toxic and both are infinitely cheaper if you prepare the solutions yourself. Acrylamide is a hassle due to the necessity to wear a mask and weigh the neurotoxic components in the hood. However, you may have to prepare the stock solution once every 2–3 months, thereby saving three times the cost. Phenol, depending on the ground state where you start, may have to be redistilled and then requires further processing, extraction, and clean-up time. This process can take a lot of time and in the end may not save you a considerable amount of money from buying it ready-to-go.

In contrast, making nontoxic reagents that you will use often may be best suited to the you-make-'em category. An easy example is competent *E. coli* bacterial cells for DNA transformation, which are easy to make. While the quality or efficiency of such products can vary between the you-make-'em and the commercial best ready-to-go versions, most experiments require the lower grades/efficiencies. At $10-20 per experiment, the costs can sky rocket over the course of a year. Preparing a large batch, aliquoting, and storing the cells could save you thousands without sacrificing your experiments.

Thus, time, money, danger, and of course, ready-to-go are the considerations in this decision process. If you do not need a lot of a reagent, then the time saved in buying it ready-to-go will be better spent on the experiment rather than on the preparation. The advice is simple:

- Buy what you need.
- Buy the quantities that you need.
- Evaluate the cost-effectiveness of kits vs. do-it-yourself.
- Budget the items so that your start-up funds last the length of time they are designed to last.

> ▪ **Important note**: Do not rely on the generosity of the neighboring lab to provide you with your reagents or small equipment that you forgot to buy. While most scientists are initially happy to share, they may be unhappy to share as you use up the last of their supplies or their funds run low. Remember, their funds are for their specific projects.

Extra Money

If you find that you have extra money at the end of a period that you have to "use it or lose it," you will now have an opportunity to make the funds go a long way. You should use this money for supplies that you will use over a long time. Examples include bovine serum, pipette tips and other plasticware, and stable chemicals. (You would be surprised at the deal on pipette tips that you can get if you order 100 boxes.) Don't forget the storage issue. Also, you may be able to use this extra money to buy a small piece of equipment that would make life easier, but was not in your initial budget. Don't waste it on frivolous items. Think long term!

Reagent Storage and Record Keeping

"Where did Sam put Elsa's DNA preparation?" As time goes by and your lab becomes larger, you will not be able to find anything or remember exactly what it was or how it was made (unless *you* made it of course). Therefore, it is important to have a system in place for reagents that are unique to your lab. Database programs, such as FileMaker™ Pro are easy to use, and are extraordinarily useful for keeping track of oligonucleotides, antisera, peptides, protein preparations, plasmids, cell lines, bacterial stocks, and anything else that you can think of which you want to have a long-term record. As people complete their journey through your lab, you should have them place their unique reagents in sequenced storage boxes and document the location of the reagent (see Departing Personnel, below). Of course, reagents that didn't work and have no value should probably be discarded at this time, with a corresponding mark in their notebooks.

> ▪ Establishing a long-term database at the beginning of your career is critical to having an accurate record of your reagents and being able to find them 10 years later.

Six databases are included on the CD that accompanies this book. These include a Freezer Storage System file (**Appendix 4-3**), a Peptide database (**Appendix 4-4**), a Plasmid database file (**Appendix 4-5**), a Reagent Inventory System database (**Appendix 4-6**), an Oligonucleotide database file (**Appendix 4-7**), and an Antisera database (**Appendix 4-8**).

Each database is in FileMaker Pro v 7.0. All of these were developed and used in the author's lab, some for over 13 years. The value of these electronic databases is that when you need to find something, you can at least have a starting place to start looking. The databases can be maintained by everyone in the lab. A work-study student can help with inventory and data. Don't forget to back up the databases on a regular basis.

Departing Personnel

Two months after your favorite student has graduated and gone on to his Big-Time postdoctoral position, you will spend several hours combing through his refrigerator stocks looking for that last plasmid that he made. Unfortunately, losing reagents and data generated by students, fellows, and employees who have left the lab is common. Use of the databases like those described in the above section or some that you create will alleviate some of the problems with reagents, but not necessarily data. In fact, the amount of information that each individual creates in your lab will be overwhelming for anyone but that person to reconstruct, and that individual will only remember it for a short time after he or she leaves.

      The only solution to this problem is to have everyone who is leaving spend time organizing his or her data. A "Departure Checklist" is provided as **Appendix 4-9**. The aim of the checklist is to have the exiting individual gather his or her information, document where the information is, fill in the database entries that he or she has ignored, and provide you with a notebook organizing this information. It takes time to use this checklist, but it has been very effective. After they have completed the tasks listed (as well as others that you may have), discuss the list, the master notebooks, etc. Visually inspect the boxes where they have stored their reagents. Make sure that they have condensed their materials and thrown away junk. You may actually want to tell them what you think "junk" might be.

POLICIES

Having established laboratory policies in place when the first or second person arrives in your laboratory makes it easy to enforce the policies as your lab grows, as others in the lab will know the policies and do the enforcing for you. A number of common lab policies are discussed below. For the most part, these policies concern productivity. Enforcement of any policy only becomes an issue when someone is unproductive or taking advantage of a your easygoing, laid back, and loving personality.

Notebooks

The laboratory notebook is the official documentation system of your research. Specific guidelines and the length of time that you must keep your notebooks may be in place at your institution. You should establish the format that you like with each new individual in your lab when he or she arrives. Ideally, the pages should be numbered and a table of contents should appear at the beginning of each book. Electronic notebooks are becoming more widely used as spreadsheets and databases are developed. If possible, a hard copy of this information should be generated and stored. If not properly annotated, dated, and numbered, items that can come out of or cannot be stored in notebooks, such as autoradiographs, are easily misplaced. Improper annotation prevents easy confirmation with the original experiment after everyone who was involved in that work has left the lab but you. Dating and witnessing (actually signing) notebook pages may be a good policy in case you are fortunate enough to discover something that is worth patenting.

Protocol Drift

Protocol drift (*verb*, a procedure that becomes altered as it is passed down through the personnel of your laboratory) becomes an issue when a perfectly good procedure is altered to the point that it no longer works. Because this occurs all the time, you may want to have established written protocols for the assays and procedures that you rely on the most, as well as those that are unique to your laboratory. You may also want to establish a policy that makes altering a lab protocol a decision made at a group meeting that requires a demonstration that the alteration is truly an improvement.

Ortholologous Protocols

(*noun*, problems arising in a lab due to the use of a wide variety of protocols to accomplish the same goal, usually accompanied by variation in quality of data or reagents.) As laboratories get larger, the ability to share reagents and be assured that everything is being done the same way becomes problematic. For example, laboratories can prepare plasmid DNA by at least 6 different procedures. For the most part, these common simple procedures may not affect the overall productivity of the lab, unless the different preparations have different levels of quality that affect the results. At this point, you may have to choose one or two procedures to make sure that you can troubleshoot experiments that do not work.

Lab Computers

Lab computers are essential for efficient data collection, data storage, and data analysis, and are tools to search and read the literature for new ideas and interpreting results. Your lab personnel will find computers essential for all of these activities and then some. Computers can also be a time sink for the undisciplined. Surfing the net, emailing friends, and day trading will use up valuable time that you and your staff don't have. For you, the young Academic Scientist, self-control is a must. For your lab staff, you may have to insist that there are no games installed on the computer or that email time is on their time not yours. No one can argue that he or she should be spending your money communicating with his or her high school buddies about last weekend's football game. Because this can get out of hand, make sure that everyone is aware of your thoughts on these issues. Sometimes the issue involves only one person who is losing focus and productivity. Be prepared to set up rules for personal computer use. You may need to limit personal use for that individual or more broadly to the lab. One rule could be that all non-business computer use must be done before 9 AM or after 6 PM. While it's hard to enforce, it is obvious when the rule is being broken as the computer monitor blinks as you enter the room. Computer peripherals may also be used for personal use. For example, downloading MP3 files and creating the new Conway Twitty hit CD on that fancy CD burner may be important to you, but remember that monkey see, monkey do.

Excessive Telephone Use

The telephone is a disruptive devise. While senior investigators spend a good portion of their days on the phone, constant interruptions in the lab can ruin good experiments. Additionally, the one being interrupted is not always the one who the phone call is for. If your lab complains that Pat Talktician is receiving a zillion calls a day, it is time to step in and tell Pat that he/she must reduce the number of calls as it is disruptive to others and to his/her productivity. Cell phones may reduce the disruption to others, but not to Pat.

The Radio

The radio can be a calming or inciting influence in the lab. For some, it is clearly necessary when they are processing a large number of samples and they don't need to think very much. For others, it is necessary background noise that allows them to be isolated in the large room. Yet others do not want to listen to NPR or the Stones, but prefer ol' Conway. While choice is something that is important, volume can be the determining factor. Lower volumes are less intrusive and may even allow Conway to be played (maybe not). Get the lab to agree on a station or times for stations if there are arguments. Personal radios/walkpersons/MP3 players, or iPods solve many problems, but tend to isolate people from each other.

Other Distractions

Because the lab will become the home away from home to your students and postdocs, it is often the case that much of their personal lives will enter into the daily routine. This may include reading the newspaper, betting on football games, dating, and etc. Each of these situations presents opportunities for the undisciplined to waste a lot of their time and therefore your time. If productivity falls, it is time to make some changes to restrict these activities. Declarations, ordering, and ranting are by far the most effective ways to state your view. However, this approach will do absolutely nothing for your lab's morale, and your lab is likely to miss the point of what you were saying. If you explain to the group that it is necessary to boost their productivity, it is for the good of the lab, or that we need to focus to beat old Dr. Scooptcha at Higher-Ranked U. to the next publication, they are likely to understand and comply.

Another method to boost productivity despite distractions is to do the "bench-warp shuffle" and move people to new benches or work areas. This is an extreme measure, but it works well. You can even use it subtly when a new person joins your group, by offering a new bench to someone already in your lab. The bench-warp shuffle is an easy way to separate two individuals who do not get along and who create a steamy environment in your lab.

> ■ Lab management is a difficult task that most of us were not trained to deal with. Communication is key to having a good relationship with your staff. Do not ignore problems. Develop your lab management plan early in your career, and when problems arise, consult with your senior colleagues and your departmental business manager.

CHAPTER 5

FACULTY CITIZENSHIP

"Someone from your lab broke the GammaZoid, so you and your lab can't use my equipment ever again." When a senior faculty member greets you with this complaint on a Monday morning, you've got both a problem and an opportunity. The problem is, of course, the fact that a senior colleague is mad at you. The opportunity lies in the way you resolve the crisis. The proper handling of such seemingly non-research-related matters is important for your ability to blend into the framework of your department and to allow you to enjoy your colleagues' charming personalities. This chapter is designed to provide some guidance toward being a contributing citizen of your department and institution. The specific issues discussed are:

- Getting along with your colleagues
- Getting along with your chair
- The 80% rule
- Choosing a mentor
- How to say NO
- You and committees

GETTING ALONG

In academic science, being a research and teaching wiz are necessary but not sufficient to help you become a valued member of your department. You don't need to strive to be the most popular faculty member, because that is irrelevant to your scholarly value to the department. You can continue to be your unconventional self, but as a new faculty member you must figure out how to earn the respect of your senior colleagues. Remember, you are the New Kid on the Block, and you want to live happily in the neighborhood for a long time.

The Scene

In the scene described above, a senior faculty member greets you with the accusation that your lab staff broke an important piece of his lab's equipment. This is a common problem in research departments, and one that you will encounter sooner or later. You have lots of choices in how to react to your colleague, and most of the choices are wrong. Using this confrontation as the backdrop, take the following multiple-choice test about how you would probably respond to this complaint:

1. In response to the senior colleague's accusation about the broken equipment, you immediately say...
   a. "Huh? What? I haven't had my coffee yet and I can't talk about anything until I do."
   b. "It wasn't my lab. It must have been someone in Dr. Kneematoad's lab that broke the GammaZoid. Go give him a hard time!"
   c. "I'm sorry about the GammaZoid. It's an important piece of equipment for your lab and mine. I'll talk to my lab first thing this morning and I'll get back to before lunch to talk about this problem."
2. When you go in the lab to talk to your lab members, you ...
   a. Forget to say anything about the broken GammaZoid.
   b. Find out that someone from your lab actually did break the GammaZoid and berate him at length and in high volume in front of the other members of the lab.
   c. Find out that someone from your lab broke the GammaZoid and determine how and why it happened so that you can keep him from breaking anything else.
3. You verify the fact that the graduate student with the curly red-hair from your lab broke the GammaZoid. You then address the problem with Dr. Rekinwith by ...
   a. Ignoring it, because you don't have any idea how to handle this and hope that Dr R. will forget about you and the fact that you are a member of the department.
   b. Forcing your graduate student to deal with Dr. R. himself.
   c. Talking directly to Dr. Rekinwith, apologizing for the problem yourself without pinning total blame on the student, and discussing the options for getting the GammaZoid repaired (along with your lab's reputation).

How did you score? You may have inclinations toward the a or b answers, and may find some of these responses to be OK — even though you realize they are self-defeating. But, if you accept the clearly reasonable answer to each question and recognize that these responses represent good citizen behavior, then you are certainly capable of responding this way — although you may *want* to be outrageous or indignant. Controlling your inclinations and doing the right thing will be much more productive for you and your career.

It is important to recognize that you and your lab members are highly visible members of the department. Your research group will take on a collective "personality" that totally accrues to you, the faculty member. Keep in mind that your colleagues informally evaluate the way you handle problem situations and departmental "politics." Even though these evaluations are not necessarily documented for your eventual tenure decision, they do become part of your work profile that can either make or break your reputation in the department and determine how much any faculty member may wish to go to bat for you when the time comes. So while you are working on your research, work on your Faculty Citizenship Quotient, FCQ.

The measurement of your FCQ is based on the way you conduct your interactions with your colleagues, students and staff. Some of the components of good citizenship are:

> - Using good judgment in "choosing your battles" when you encounter disagreements
> - Doing your share of department and institutional responsibilities
> - Being accountable for the work performance of your graduate students and postdocs, and for the behavior of your lab members
> - Being consistent in the way you treat your staff and the way you interact with your colleagues
> - Holding your temper, even when an outburst seems justifiable

Choose Your Battles!

As you go through your academic life, you will find that you and your colleagues will not always be on the same side of the fence when it comes to whom to hire, what order to give a series of lectures, how the seminar program is scheduled, or what day it is on, etc. Which, if any of these, is so important that you must win the argument? This is the critical decision you must make. While you may be able to make the best cogent argument against Dr. Dewit Miweigh's plan for the faculty and student retreat, it is unlikely that his plan or yours will really matter to the success of the retreat. Dr. Miweigh, however, will remember that you opposed him on something that he felt strongly about. Because the retreat does not really affect your ability to do science and succeed in your life as an academician, blocking Dr. Miweigh's plan is foolish. On the other hand, the new departmental space policy being presented by Dr. Miweigh to move the shared GammaZoid to a different floor will affect your scholarly pursuits. Thus, in this case you will need to think of an alternative to his plan and present that instead. The alternative should be something that both you and Dr. Miweigh will be happy with. Of course it is possible that Dr. Miweigh did not even consider the fact that you would need the GammaZoid.

> - The bottom line is to choose the battles that have long-term consequences that are important to you, ignore those that do not affect your career or the treatment of your staff, and enjoy the ramblings of your elder professors.

Do Your Share

You must do your share. Departments run on the academic contributions of all the faculty. A faculty member who does not help out has less of a voice when it comes time to change things. However, as a beginning faculty member you must be careful about taking on too many obligations. There are tasks that you will like and those you will loathe. Look and listen to what is going on and try to get involved in those duties that suit your personality and avoid those that do not. Refer to the section on Saying NO for ways to sidestep the jobs that you would prefer not to do.

Be Accountable

You are responsible for your lab members' behavior at work. The point here is that academic institutions are serious workplaces. You and your lab must be professional when it comes to behavior and performance.

Be Consistent

Your actions and views, while they are allowed to change, need to be fair. This is important in the way that you treat your staff and the students in the programs in which you participate. You need to "do the right thing" as consistently as possible.

Back Off!

You must keep your composure. Sure, it can be strangely satisfying to yell at the secretary in the front office because he mailed the draft copy of your grant off to the NIH, but will that get your grant back in your hands? Carefully phrased words will move things along faster than provoking everyone. Most things can be repaired in a timely manner, including getting your grant back from the NIH without a problem (personal experience). If problems persist, or if you feel that a particular incident with the individual needs fixin', then talk with your departmental business manager or your chair about the issue.

As far as students and postdocs go, unless you are the best of the best, losing your temper with your lab staff is a sure way to keep others from joining your group. Professionals can perform consistently and with equanimity even under duress, including the waning hours as your grant deadline is approaching. The reputation that you develop will be with you, always.

> ▪ You can raise your FCQ through such time-honored human relations techniques as treating people the way you would like to be treated, and modeling your behavior after that of someone you know who clearly gets along well with colleagues, students, and staff. If you recognize that you have a problem getting along with people, then do your best to avoid getting into difficult situations.

> **Survey Says**: A tenured, full Professor responded, "The best advice I got [as a junior faculty member] was to focus on my work, and stay away from institutional politics and politicking."

SHOWING UP!

Woody Allen's assertion that "showing up is 80% of life" certainly applies to the academic work environment. As noted earlier in this chapter, your faculty colleagues observe your work habits throughout the years leading up to your promotion/tenure decision. If you are not around to be observed, your absence will be remembered and talked about by the other faculty, as well as by the students and the staff. It is difficult to insist that your lab staff work late or come in over the weekend if you are never there. Monkey see, monkey do applies to the work environment. If you want your lab to show up, you must as well.

One of the benefits of academia is that you set your own hours. While many of us are nocturnal and loathe crawling out of bed at 7 to get in by 9 AM, working the day shift has its inherent advantages and disadvantages. The obvious advantage is that you are there during the day when most of the action occurs. Your colleagues, students, and staff can easily find you and talk with you about their science and other issues. This is, of course, also the disadvantage, because everyone can find you and disrupt your writing or experimentation. However, in most cases being available is part of the game. If you are a night person, you may have to alter your clock a little to arrive in the late morning so that you are available for most meetings and seminars. Morning people have it easy. They arrive before anyone else gets in and complete most of their work in the peace and stillness of the sunrise. However, they don't really know who Jay Leno or David Letterman are or why some think that they are amusing.

Showing Up at Seminars

Seminars and research presentations are the forums for scientific exchange in most institutions. Local research-in-progress meetings are short informal sessions that most departments or graduate programs organize. It is important to become a regular participant in such programs for four reasons. The first is that research-in-progress type meetings by the students and postdocs let you know at a primary level about what is going on in the labs of your colleagues. While some of the presentations may be painful, others will impress you and you may find the perfect collaborator two floors up from your lab. Second, your ability to participate in the discussions at these meetings will highlight your brilliance, something that you want everyone to see. Of course, if you are not so brilliant, or just plain obnoxious, everyone will see that too. Third, the senior faculty will see you as a teammate and a player. In time you will become part of "the group." If your tenure decision comes down to the wire, being a team player may push you over the top. Lastly, it's what science is about and it's fun.

Most institutions sponsor numerous outside speaker seminars. Invited speaker seminar series serve several purposes that you should consider important. The first and obvious point is that by attending these seminars you will be brought up to date with a field or an area and hopefully after listening for an hour you may be able to incorporate one of the experiments into your own work. The second is that because most outside speakers spend a good portion of the day talking with the faculty, there is an opportunity to advertise yourself to them. If Dr. Hotshot is coming through your school to give a talk, try to get on her visitation list. You may even get to go out to a fancy restaurant for dinner. When she arrives at your office, be prepared to show her some cool data. Be efficient in your setup and presentation. It is important to keep your eye on the clock so that the speaker won't be late to her next appointment. If the speaker is late in getting to your office, call the next person down the list and negotiate when you will bring her to him. This way your colleague won't blame you for eating up his time with Dr. Hotshot.

If you really think that you need to meet Dr. H., consider finding out how to get funds to invite her yourself and be the host. If you are the host, you should know that during Dr. Hotshot's visit you will be busy making sure that the day goes according to plan and that everyone who wanted to see Dr. H. had the opportunity. So plan to catch up on your reading during that day. Many junior faculty are asked to organize their departmental seminar program. This is a good thing. Here is your opportunity to meet the people in your broad area of interest and impress them with your brilliance and charming personality. After all, you really never know who reviews your grants and papers, or who it is that sends in a positive letter for your promotion to associate professor with tenure.

Showing Up for Class

This is a no-brainer. Be there, be on time, be prepared, and be ready to go!

Showing Up for Committee Meetings

We all have missed a few in our day. And some of us are perpetually late. Yes, we blame it on the e-calendar, the secretary, but mostly we should be blaming it on ourselves. If you keep your colleagues waiting, they will not be happy with you. They will remember it and if they are clever, they will schedule a 2:30 PM meeting for 2:00, so that you will show up on time. (No, no one has thought to use this with me yet.)

Lunch

"Lunch? I don't have time for lunch; I have my promotion review in five years." Doin' lunch with your colleagues is the easiest way to get the scoop on all that is going on. Your colleagues are likely to discuss who is doing what to whom, where the new space is, problems they are facing with their grad student with the curly red-hair, the new chairman, and maybe what they think it takes to get promoted to Associate Professor. At lunch, you can get the "skinny" on the department and the institution in a short time frame. You may even become close friends with your lunch colleagues. There is one downside, however. If you spend two hours at lunch everyday, you will never get anything done. So, limit your time for lunch, but make the most of it.

> ▪ The bottom line is that there are far-reaching benefits to the informal relationships you develop with your colleagues and the scientists who pass through your department. Don't miss out on these goodies by not being around.

GETTING ALONG WITH YOUR CHAIR

The chair of the department recruited you and is, at least initially, glad that you joined the department. Do your best to keep her in this frame of mind, even after you discover that she has some quirks that bother you. The chair's shortcomings are not your concern. The chair will most likely continue to see only the best in you until you do something to disillusion her. Ideally, by the time she realizes that you aren't perfect, you should have your research program running well enough that it won't matter to her.

Your job is to focus on your research, teaching, and service commitments. The chair's job is to evaluate your viability as a faculty member of her department. So help her do her job by keeping her informed about your progress, even if she doesn't ask you about it. Take the initiative by scheduling periodic meetings with the chair specifically to tell her "how you are doing." During the first year, you should probably schedule such a meeting every 4 months or so, bringing with you a document that shows your goals and accomplishments. These accomplishments include your progress in establishing your lab, hiring personnel, applying for research grants, and mapping out your research plans for the coming months. An example is provided in **Appendix 5-1**. Review these accomplishments in detail with the chair and elicit her comments about the direction you are taking with your lab. It's good to find out if she thinks you are off base as soon as possible so that you can discuss this with her and learn about her view of how junior faculty should establish themselves. Be open to her professional advice and use these discussions as opportunities for learning the ropes of the department and the institution in addition to your providing information about your progress.

Keep in mind that "being open to advice" from the chair might lead to the difficult situation of your not agreeing with the advice she gives you. As an Assistant Professor, the double-edged sword of seeking advice is that you certainly *need* advice, but if you don't *heed* the advice you are given and the advice turns out to be correct, your colleagues will remember what they told you. Because you are striving to succeed and not annoy people or develop a reputation for not listening, be careful how you ask for advice from your chair and your colleagues/mentors. Avoid asking, "What should I do about ..." and instead discuss issues with the chair and other senior faculty by asking for their experience with similar situations. This allows you to gain knowledge from their perspective while keeping both of you from being cornered into giving or receiving "The Word" on exactly how you should handle the issue at hand. All of this is tricky, so be careful how you approach your requests for guidance. But by all means, do take the opportunity to learn from the experience of your chair and your colleagues.

Finally, getting along with the chair of your department is not just related to *your* career. Because the chairs of most research departments are usually active scientists themselves, you can take the opportunity to talk to the chair about her own research plans and the plans of her postdocs and students. She is a colleague in addition to being your boss.

CHOOSING A MENTOR

Among the advantages of working in academic science is the fact that there are lots of senior colleagues around who have been through the steps involved in establishing successful research programs. The majority of your colleagues want you to succeed in your research career and are often very happy to help you as an advisor and "mentor." The prevalence of informal "mentoring programs," as well as more systematized programs of assigning mentors to junior faculty, has increased in higher education in the past decade as the value of having mentors has been recognized. In fact, several top research institutions have established formal mentor programs for their junior faculty. One measure of the importance of mentors is the assessment of faculty members who missed the opportunity to have them.

> **Survey Says**: A tenured Associate Professor — "It is very important to get mentors early. My department was not very good about mentoring their junior faculty, and it took me much longer to learn the ropes. I made all the typical mistakes because no one bothered to take the time to tell me what to do and what not to do."

If your department has a formal mentoring program, talk to other junior faculty about their experiences with their mentors. Then, talk to some of the tenured faculty who you believe might be able to help you either by offering their commitment to meet with you during your tenure-track years, or by providing you with advice about which faculty members would be good candidates to fill this role with you. (You also need to know whom to avoid.)

- An **appropriate** mentor has the following attributes:
  Tenured rank, preferably in your department
  A funded, active research program
  Is not your scientific competitor
- A **good** mentor has all of the above attributes, plus
  The respect of his or her colleagues
  Former "mentees" who have attained tenure
  Has served on study section
  Has served on an editorial board of a journal or is constantly
    reviewing papers
- A **really, *really* good** mentor also has experience with the tenure
  and promotions process of the school or university and therefore
  knows the essentials of successful tenure portfolios at your
  institution

Take your time picking a mentor because you'll be working together for several years. Establish the parameters of your working relationship by determining whether or not you will schedule periodic meetings on a regular basis, or if you will simply call upon your mentor when you want help. It is primarily your responsibility to maintain the relationship by staying in contact with your mentor. Use your mentor to read and discuss your grants and papers. However, as discussed in *Chapter 3*, if you do not give him enough time to read your grants or papers critically, you will not benefit from his help.

If your department doesn't have a formal mentoring program, speak with your chair or senior colleagues about the possibility of implementing one. Alternatively, you may find a colleague who will agree to informally mentor your career and provide you with the advice that you seek. You may even want to have more than one mentor. This way you can spread your requests for help around or bounce the ideas of each of your mentors. The bottom line: Don't be shy about seeking help and advice!

How to Say "No"

"Here's a terrific opportunity for you, Kid. Take my place on the University Parking Committee. They only meet every other week for 3 hours, and it's just a two-year appointment. I think this will look good on your tenure application." You realize, of course, that you should not take this assignment. But how do you turn down the full Professor who offers you this so-called opportunity? Here are some suggested responses:

- "Thank you, Dr. Kneematoad. Let me take a couple of days to look at my calendar and to talk to my mentor Dr. Pat I. Ence to find out what she thinks about my doing this."

- "Gee, Dr. Kneematoad. That sounds interesting, but I've got October 1 and February 1 grant deadlines, so I really can't commit to anything that takes me out of the lab and away from my computer until after I get this grant funded."

- "Wow, that would be a great honor, but I collect so many parking tickets on this campus that I'm sure that I am disqualified from serving on the Parking Committee!"

- "My laundry is ready for pickup, I will get back to you."

Because in your Assistant Professor years you are trying to focus your efforts on your research program, you should only accept assignments that are science related. The key to saying NO to service commitments is to avoid being simply dismissive of offers to provide service for your institution. You can use delaying tactics, rely on your mentor to advise against your taking certain assignments, or deflect the offer by changing the subject. Do NOT simply stand there and acquiesce just because you can't think of anything to say. And, certainly don't accept an inappropriate service assignment because you're "honored" to be asked.

Also be prepared with responses to other requests that you need to avoid. These may include situations such as an undergraduate student who asks to work in your lab on a project unrelated to your research, or a colleague who continually asks you to perform a laboratory procedure for him. Because you have limited time to establish your career, you need to assert yourself and explain — non-defensively — the reason that you cannot comply with any requests that infringe on your time with little or no benefit to you. Remember that being assertive does *not* mean being rude to students or colleagues — or to anyone else who asks you to help them out.

Although it sometimes seems easier to go along with simple requests than to come up with a suitable reply that gets you off the hook, you should prepare yourself for these situations. The point is that you have to focus on your own work and avoid time-consuming responsibilities that depart from your professional objectives. And you have to remain a good Faculty Citizen throughout.

COMMITTEES AND OTHER MEETINGS

While *Chapter 9* will focus on running a committee so that it is efficient and productive, this section will discuss what is expected of faculty who come to committee meetings and provide some advice on what not to do; i.e., committee dos and don'ts. Most academic institutions run on the premise of a democracy with groups making the decisions for the whole. There are limits to what can be done and what should be done, however. These limits are established by higher-level administrators, such as your chair or dean. In general the purpose of a committee should be to discuss a plan of action with short- and long-term goals. It is in these meetings that you need to choose your battles. If Dr. Dewit Miweigh is bullying the committee to go his way, then as a junior faculty member it is not your responsibility to stop him. If your chair is part of the committee, telling your chair that her plan is foolish is also not a good thing. From these comments, it seems that junior faculty should be seen but not heard. This too is not completely true. Like most other things, there is a sensible balance in how you should participate. While you may intend to retire as a tenured full professor at Research Utopia University, you are not yet a permanent member of the faculty, and therefore, your views are only suggestions. The advice here is simple:

> - Choose your battles
> - Think before you speak
> - Do not insult anyone or anything
> - Be polite regardless of your stance

There are additional points that will help as well. If there is an assignment that goes with the committee, get it done on time (not vital but imperative). If you do object to the way something is done because you *have* in fact seen it at your former institution, point this out and give the rationale for your view. However, be careful of always saying "at Hahvaad we did it this way," as it will get old to the listeners and someone may eventually say "this ain't Hahvaad."

> - As a junior faculty member, strive to establish yourself as a leader and respected colleague in your institution. Remember to choose your battles carefully and always maintain your goal of being a successful Academic Scientist.

PART II

DOWN THE STRETCH

The chapters in Part II present approaches to organizing, planning, and achieving excellence in the three criteria that define an Academic Scientist: Scholarship, Teaching and Mentoring, and Service.

CHAPTER 6

BEING A SCHOLAR

"I got funded; ergo, I'm a scholar, ain't I?" Well, maybe, maybe not. A scholar is an individual who is an expert in a subject, which in this case is some field of science. Scholarly achievement is the most important criterion used to evaluate an Academic Scientist's career. While subjective in nature, scholarly achievements can be documented in a number of objective ways such that non-experts in your field can evaluate you. The ways include publication in scientific journals, continuous procurement of extramural funding, appointments to positions where your scientific judgment is required, inventions and patents, invitations to present your work at other institutions or at scientific meetings, etc. It is important to consider the point that once you take your new position as an independent investigator your scholarly achievement slate is wiped clean, at least as far as your new institution is concerned. You now have no papers, no grants, no invitations, etc. Oy Vay! So where do you begin? This chapter will discuss organizational approaches to scholarly activities. These will include:

- Strategies to organize your research program
- Preparing and submitting manuscripts
- Responding to critiques
- Exercising your scientific judgment

ORGANIZING YOUR RESEARCH PROGRAM

The organization of your research program will evolve as you gain experience in supervising your staff, gain or lose funding, or alter the number of personnel that you supervise. As presented in *Chapter 4*, managing your staff not only means managing the way they function in your lab but establishing the long- and short-term goals of their projects. Now that you have written your RO1 grant, you actually have an outline of the different experiments, a potential order in which to do them, and you may have already assigned some students to work on various aspects of each of the aims. You should now ask yourself "is this the best way?" You should ask yourself this question at least once a year. The reason to do this is that the ability of each staff member to make progress on a scientific aim is different. Finding the right fit for someone's talent is critical to the progress of your program and your overall success. You do not want to give the hottest project that you have to someone who can't do it or is too inexperienced to perform at the level that it needs to succeed. If you do, you will not be successful. So, how should you go about assigning projects?

Strategies

There are at least three distinct strategies that you can use in assigning projects: Solo, Mini Group, and Assay. The strategies may also be combined. Each has its advantages and disadvantages.

Solo

In the "solo" strategy each person in your lab has his or her own, non-overlapping project. This strategy assumes that these people will not be publishing together. Graduate students and postdocs tend to like this strategy because they have something that is theirs and theirs alone. However, when they get stuck on a technique or a procedure, the project gets stuck too. If you choose to use this strategy, the more difficult the project, the more experience the person needs to make it succeed. If your lab has relatively inexperienced hands, then the more difficult assays and procedures need to be performed by you at least until you can train someone to take over this part of the project.

Each person's project must have both long- and short-term goals. Even with the solo method, beginning graduate students should be assigned very short-term goals and projects that they can complete over a 1- to 2-month period. This will allow you to assess their capabilities and to monitor their progress. As they gain experience, the goals can be extended. If you have hired a postdoc, his or her goals can be more long term, like an entire aim of a grant. However, even here you should make sure that your postdocs have some short-term goals so that their progress can be monitored.

Mini Group

The second strategy is the "mini-group" strategy in which a small number of people contribute to both short- and long-term goals. This strategy puts more hands on a single goal and can foster quicker progress. Many labs will adopt this strategy if they feel that they have discovered something hot and a lot of work needs to be done quickly so that they do not get scooped. An example of this strategy might be that you have identified a new gene and the molecular and cellular characterizations of the gene are required before you can submit to your favorite red-hot journal. You might have one person find out where the protein product resides in cells, another may characterize its expression, and the third may make a variety of mutants in a critical domain of the protein. If one person were working on the project, it would take a year to do the three outlined goals. If three people are working in this manner, then each can proceed simultaneously and you will be submitting your paper sooner rather than later. The other advantage of this strategy is that your lab staff will interact very closely and the excitement on any one project will hopefully be contagious. You might now ask the question, why wait to employ this strategy? A major drawback to this strategy is authorship and project ownership. Who's on first? Authorship issues are discussed below. Every project needs a lead person who is in charge of making sure that the i's are crossed and the t's dotted. It has to be clear who is the leader. Generally, the person who has made the initial discovery is the leader. That was easy. But this is not always the case. To avoid serious problems in this regard it is important to let everyone know what their role is on the project. If you employ this strategy all the time, you will establish several projects within the Mini Group, and each member of the group can have the potential to lead a project and be a first author. This strategy is a nice way to pair a new graduate student with a senior postdoc.

Assay

In the "assay" strategy, the workload is organized by technologies and assays, and is most successful when used with technical support personnel. This strategy allows each person to become a true expert in a particular assay. It also allows you to easily supervise everyone's workload and progress, modify their experiments, and to analyze their data. This is also the easiest way to set up your lab and minimize the amount of specific training that you have to do. However, the drawbacks to this system need to be mentioned as well. Research lab staff like to have their own projects. This is especially true of postdoctoral fellows and graduate students. This approach also has the potential for brief delays in data collection if only one person knows how to perform a series of experiments and for some reason leaves the lab before the work is complete. Thus, cross-training is important, even if this system is applied.

Most labs will use a combination of all three strategies to achieve their goals. Importantly, you will need to decide how YOU will fit into the strategy that you employ. When starting out, you should take on the most important project, train and organize your support staff using the assay strategy, and be sure to get your personal project moving. At the same time, you will need to employ the other two strategies as the size of your laboratory increases.

Punting

Although often reserved for football, punting should also be used when a project is deep in the hole and you need to move on (4th and very long). The key issues in deciding to punt or not are: why, when, and how.

Why?

While we are in love with all of our ideas, some will not work for unknown reasons, or will not work with the person doing them. The hypothesis might even be wrong, too. Because a 5-year grant is renewed in the 4th year of funding and you may have only 5.5 years to prepare your promotion or tenure dossier, making the decision to punt at the appropriate time critical to your success and ultimately to your students'. In many cases, it is not important to figure out why the project or procedure did not work. This sounds like harsh advice and a quitter's attitude. It is — I quit. Now let's move on and enjoy science.

When?

Unless you can step in and provide the next step, it is time to change a student's or postdoc's project when progress has been slow or nonexistent for 4–6 months. This assumes that the experiments do not take 4 months to complete. This also assumes that you have explored most options and that the student/postdoc has made a serious effort on the project. If doing another set of assays that will take a few weeks will lead you to success, do not punt. However, if the first three-week extension was followed by a second and third... You get the idea: PUNT!

How?

How you change the project depends on the goals that were initially put forth. If the staff member has "good hands," then you may have to alter the approaches that you are taking on your project. Remember the alternative sections in your grant (*Chapter 3*)? Reread this section of your grant and refer back to the critique/summary statement, as it is possible that the reviewers may have provided you with some tips. If one aim of the project is dependent on completion of this set of experiments, then you may have to alter your strategy and goals. This should not be taken lightly and a brainstorming session with some of your colleagues may be a good approach to coming up with new ideas. If the aims of your program are not dependent on this aspect of the project, then moving on to the next aim/subaim may be the best move.

> ▪ Regardless of how you change your direction, spending time organizing a plan will quicken your recovery and future success. Outline the plan on paper and use it.

Oh, by the way, we have all punted, often.

Our *Next Wave* article, **The Red Herring**, reproduced in *The Extras Section*, discusses following a bad scientific lead and organizing a recovery in more detail than in this chapter.

PREPARING AND SUBMITTING MANUSCRIPTS

I must publish, I must publish, I must, I must... You must communicate your ideas to the scientific community. This is how the science world knows that you exist, and it is how they will evaluate your career. The issues are simple. When do I have enough data to publish a paper? How do I organize my paper so that it will be accepted? Who should write it? Which journal should I send it to? The order of these questions is also the order in which you should think about your work.

> **Survey Says**: A Tenured Professor — "If your research productivity is good enough in quality and quantity, as evidenced by your independent publication record, nothing else matters. If your research productivity is borderline, excellence in teaching can swing the [promotion/tenure] decision [in your favor]. **If your research productivity is inadequate, nothing can compensate.**"

When?

Determining "when" a series of experiments is ready for publication is the hardest of the four questions raised above. Sometimes it is obvious because all of the experiments have gone exactly as planned and shed new light on your area of research. Other times, you have collected lots of data, some of the results flow in a continuous theme, but some do not. So, you need to decide when enough is enough. As you gain experience, choosing the time will become easier. Below is a list of issues that you should consider in deciding if you have enough data to submit a manuscript.

> **Survey Says**: A Tenured Professor — "Get your research articles out in a timely manner. One a year for 4 or 5 years is better than 4 or 5 the year before promotion."

The Minimal Publishable Unit (MPU)

Most manuscripts make a single point with many of the experiments serving as controls to support the initial or important observation(s). The more complex the interpretation, the greater the number of experiments one needs to support the conclusions. Thus, an MPU is a manuscript that contains the smallest number of experiments that tests the hypothesis or answers questions posed. At the beginning of your career, focusing on what might be the MPU could serve to help you and your young lab produce publication-ready data and help guide you towards your first publication(s). Thinking about the MPU may prevent the Magnum Opus Syndrome of waiting until the entire system is solved before publishing. Magnum Opus strategies can result in getting scooped, or worse, not being promoted because you did not submit your work in time to allow for the relatively long period of time that journals spend before they accept a paper.

Hot Data

It is always best to prepare and submit your manuscripts when your data are hot, fresh, and of course correct. This usually coincides with the discoveries being novel and exciting. Waiting on hot data for more than a year is not a good approach as someone else may scoop you during that time period. Additionally, your overall enthusiasm (and potentially a reviewer's) for your results will be diminished.

Completing a Short Story

It is time to write up your work when you have answered a short series of questions pertaining to your research area. This could be a sub-aim of your grant or an entire aim. The data should support your conclusions and be convincing. Importantly, the experiments describing your conclusions should be complete and well controlled. How complete is complete? If one experiment leads to a conclusion then one could say that the data suggest that this is so. If two different experiments lead to the same conclusion then the point may be demonstrated, and a third piece of supporting evidence may be enough to prove a point.

Figure Quality

Your data must be "figure" quality. To be figure quality (i.e., ready for publication), the order of the samples should be correct, the controls included, and the overall result should be convincing to everyone. While this point seems trivial, most students/postdocs collect a lot of data that is not figure quality and the experiments have to be repeated. There also should be enough replicates of the experiment to be sure that statistics can be applied if the data are quantitated. Make it a habit to place a tab/sticky note on notebook pages in your lab that are figure quality and potentially useful for publication. If the data are electronic, organize it in folders for the paper as it is generated.

Organizing Manuscripts

Most papers have eight parts: Title, Abstract/Summary, Introduction, Methods, Results, Discussion, References, and Data. It is expected that the results will support the title and main conclusions and that the information or point of the paper will add to the knowledge base of the scientific community. Manuscripts should be well organized and written. It is not necessary to write the manuscript in the order that the experiments were performed nor is it necessary to tell the story in the order in which the experiments were conceived. You only need to tell the story. A general format/outline style worksheet is provided in **Appendix 6-1** to help with organizing and writing manuscripts.

Figures and Results

It is best to begin with the figures and the results section. In the best case, the figures should be completed at the time you start putting the manuscript together. This lets you decide what you have, the order in which they should be placed in the results section, and what additional experiments/controls you may need. Arrange the order of the figures so that there is a logical progression of ideas, questions, or hypotheses. While there is always some variation between manuscripts, Results sections typically follow a simple motif. For each figure introduce the problem at hand, describe the approach, present the result, and lastly provide a short conclusion. Repeat the motif for each figure set.

Methods

After the results are written, it is now easy to organize the methods section and describe how you did each experiment. Most well-known procedures should be referenced. However, it is important to describe everything that is unique to your procedures in this section so that the work can be repeated/verified by others.

Introduction

Most introduction sections are similar in their design. The purpose of the introduction is to provide non-experts enough background information so that they can evaluate your work. They do not need to know all that is known. Typically, basic information about the function of the system is provided first. Details of the specific area of focus are provided and sometimes the general hypothesis, conundrums, and/or paradoxes are explained. The general rationale for studying the system is sometimes provided as well. The section is concluded with a series of sentences, stating the unique/novel findings and how these findings contribute to understanding your field.

Discussion

The discussion should place the results and conclusions in the context of your field or the broader scheme of research. Most discussion sections summarize the main points or conclusions from the data in the first paragraph. Authors typically then discuss those main points in the subsequent paragraphs. They also present and/or discount alternative interpretations of the data. Models of how the system functions with the new results can also be presented and discussed. While writing the discussion is sometimes the most difficult part there are two things that you should not do. (1) Do not repeat the entire Results section. This is very boring to the reader and doesn't show your depth of knowledge or placement of your work in the field. (2) Do not over-interpret your results in this section. Critically evaluate your data and be sure to provide alternative interpretations, even if you discount them. But do not sell yourself short either.

Title, References, and Figure Legends

Everyone wants a snappy title, even book writers. But sometimes, you can't always get what you want. The best titles are concluding statements that describe the essence of your work and its novelty. A reader in your field should be able to accurately guess at what you did or have shown from your title. Shorter is always better.

Cite the references that are necessary to back up your statements. You do not have to cite all the references in your field. But if there is a cadre of reviewers who you think might end up reviewing your paper, be sure to include their important references in your paper. They will look to see if they are cited! So will you when you get a chance to review their work.

In preparing to write the figure legends for your paper, look at several articles in the journal as well as the instructions to the authors. Some journals want only a description of the experiment without an interpretation, while others want the interpretation with some description of the experiment. Be brief, but complete.

Abstract

The abstract or summary is critically important and should be written last after all of your ideas have been put into the manuscript. It should be a summary about the work and its relevance to biology/medicine. The last sentence in the abstract is perhaps the most important. Here authors usually state the grand conclusion and relevance of their work. The grander the conclusion, the more interest the article will have for the editor. Of course if the grand conclusion is overstated and is not supported by the data, this too will be judged by the editor and the reviewers.

Who Should Write the Manuscript?

Should the student/postdoc write the paper or should you as the PI write it? This is a hotly debated issue, and your colleagues will have different views on this subject. These views will vary depending on how "hot" the work is, where your colleagues are in their careers, and how aggressive they are in their own pursuit of science. There are three approaches, each with its own advantages and disadvantages.

Approach 1

You write the paper. Advantage: You have the most experience and it will take the shortest amount of time. Disadvantage: The students/postdocs will not get experience in writing. Solution to disadvantage: Have them plan, read, and edit the paper with you as you write it.

Approach 2

The student/postdoc writes the paper, and you edit it and have him enter the changes. Advantages: The student/postdoc gets writing experience. He will improve his literature base and the experience will force him to think about science in broad terms. Having him write the manuscript also gives you his point of view, which sometimes can be a surprise. Disadvantages: He may not complete the writing of the paper within a reasonable length of time or the quality of the writing may be poor. He also may not include your editorial changes, resulting in your having to spend the same effort twice. Solution to disadvantage: Give him a reasonable timeline to complete the draft. If he cannot write, then editing the manuscript with him present can often teach him how to write better. While this "House of Pain" approach of having him sit there while you read and edit may be boring to him, tell him what you are doing and why. Hopefully, he will adopt the strategy of watching and learning from the master.

Approach 3

Give your student/fellow one chance. Advantages: By giving her a chance to write the manuscript, she will get a chance to explore her writing skills, read the literature, and think in broader terms. It also allows you to take over to push the paper out the door. Disadvantage: She gets only one chance at each section. Here the House of Pain can also be used in your first editing. However, if the version that she gives you is not going to go anywhere, you should start over and have her read it along the way.

Of the approaches presented, Approach 3 fulfills some of the training aspects of being an Academic Scientist. But it also shortens the amount of time you have to wait to get your work submitted. Remember, it's Your job, Your grant, and Your career. So, if Approaches 2 and 3 are not working, go to 1.

Choosing the Journal

Journals differ by the breadth and quality of work that they publish. The higher a journal is ranked the more likely it is that papers published in that journal will be read and cited. Thus, citation ranking establishes a general ladder for journals. Journals can be categorized into four basic groups: Top-of-the-Heap; Field-Oriented Close-to-the-Top; Society-Sponsored; and Sub-Specialty journals. The more novel, paradigm shifting, and revolutionary a paper is, the more likely it will appear in the Top-of-the-Heap journals. The more narrowly focused and specialized the paper, the more likely it will be accepted into a sub-specialty journal. So, how do you decide? Consider the following about each of the journal categories presented below.

Society-Sponsored Journals

Your work should be at least at this level. These journals are supported by societies of scientists that make up a discipline. Most scientists publish extensively in these journals. These journals accept 20–40% of their submissions and provide a journal of excellence for their fields. Work published in these journals is considered important and is generally considered to be of high quality. If you publish your work in scientific society-sponsored journals, everyone in your field will know you, and your work will be read and cited.

Field-Oriented, Close-to-the-Top Journals

You should shoot for this level. These journals have a lower acceptance rate than the society journals and will demand that the data fully support the conclusions, and that the conclusions are novel enough to be worthy of their attention. The overall novelty of work published in these journals is often the same as those in the Top-of-the-Heap journals, but the topic may not be as broad or the impact of the conclusions may not be as great.

Broad-Based, Top-of-the-Heap Journals

You should reserve your best work for these journals. Of course, publishing in the top-ranked journals is best as your work will be seen by more people and is likely to be cited more frequently. However, it is very difficult to get your work accepted in these journals, as the criteria are the most stringent. Publication in these journals is often associated with conclusions that open up new avenues of research, shift paradigms, or solve biological problems that have existed for some time. Additional criteria could include the area of research that is hot that month or their competition with their neighboring journals.

Sub-Specialty Journals

This is the largest group of journals. These journals are also ranked by their citation frequency as well. Most of these journals publish good work that advances the field. However, these journals do not have the general readership of the society journal. It is also a place where your MPU may go. In short, if you have to send your manuscript there, send it.

Now that you are thinking about where to submit your work, consider the following points. If you don't ever submit your work to the higher ranked journals, your work will never be published in these journals. The advice is that if you think your work is good enough for the Top-of-the-Heap journals, send it. You can always send it somewhere else if it is rejected. Also, consider that manuscript review takes time. Thus, if you need to get your work accepted sooner, you may want to shoot for the journal in which you think your work will most likely be accepted. Your senior colleagues may be able to advise you on where to submit your work.

When reviewing a paper to be submitted to a Top-of the-Heap journal, a senior colleague remarked: "Why should I continue to read this?" His implication was that the paradigm shift and gist of the manuscript for such a journal had to be in the first few sentences; otherwise, there was no point in continuing. If you are going to submit to a TOTH journal, then you need to clearly state why you think the work belongs there.

Our *Next Wave* article **Publishing at the Top of the Heap**, reproduced in *The Extras Section*, provides additional discussion to the ramifications of journal selection and your career.

The Cover Letter to the Editor

While many journals now have electronic submission forms, it is still a good idea to write a "Rah, Rah!" letter about your paper to the editor. The format is simple, and like everything else, has 3 parts. An example is provided in **Appendix 6-2**. The first part is introducing the manuscript to the editor and pointing out the salient features that you think make your work novel and acceptable to their journal. The second part recommends a list of reviewers that you think have the expertise to review your paper. The list should include scientists in your field whom you respect. The third part should be a list of people that you do not want to review your manuscript. You can request exclusion of reviewers based on competition or personal relationships, although the latter is tricky. Editors will typically use your list of reviewers for two reasons. The first is that they don't have to generate a list themselves; and second, if they use your list, then any complaint that you may lodge later about the reviews would be invalid.

RESPONDING TO CRITIQUES

Rejected!

"Those !@#$!@#!$@#!s How can anyone possibly say that about my work?" Welcome to the ups and downs of science. Anonymous peer review is what keeps the science process honest and fundamentally sound. Unfortunately, there are some drawbacks to the system. A harsh review of your work is painful and upsetting if viewed as a personal attack of your ideas and thoughts. But, this is our business! The review is therefore only a steppingstone to revising your work and/or conclusions and resubmitting the manuscript to another journal. This is the only way to deal with peer review. Of course, if the reviewer says that the work is the best thing since an automated bread slicer, you may feel pretty good.

If your work was not perceived as better than a bread slicer, then you will have to address the comments of the review. The letter from the editor will indicate the level of enthusiasm that she has for publishing your work. In some way, shape, or form the letter will say that the work is one of the following:

- acceptable
- acceptable after minor revisions as recommended by the reviewer
- will require revision per the reviewers' comments
- will require substantial revision and re-review
- not acceptable for publication

Some journals reject everything that is not "accepted" outright by the reviewers, but leave the option to resubmit once you have thoroughly addressed the reviewers concerns. This sounds like a "reject" but is not. If you are unclear about the meaning, contact the editor who will explain what she means. If you are going to contact the editors, be sure to have a plan of what you will add or delete from the paper that could make the reviewers and the editor happy before you call. Do not call when you are angry. An editor is less likely to help you get your paper accepted if you accuse her of mishandling your manuscript.

Responding

Now that you know the status of your manuscript, it's time to prepare your response and course of action. The first thing to do is to summarize all the comments. Comments made by more than one reviewer or a reviewer and the editor should be highlighted. If these comments request more experimentation, then you will probably have to do them. If an experiment can't be done in your system, then devising an alternative experiment to get at the issue is what you must do. You should strive to address all the comments. If the list is too long, concentrate on the main points. While that sounds easy, some comments will get you mad and you may want to argue. Remember to pick your battles carefully, because you won't win them all. Successful arguments will require hard facts.

Often reviewers comment on specific interpretations that you made and request that you change these to include additional possibilities. Most of the time it is just as easy to agree and add or delete a sentence to satisfy a reviewer's concern. Because it is unlikely that that sentence will change your thesis, go with it and get your paper published.

Your response letter should highlight the reviewer's concerns that you have addressed or take issue with. Remember to be professional and stick to the point, and with only a little luck, your paper will be accepted.

One last point. We all try to guess who the person was who dunked our paper or grant. According to several editors and NIH program officials, we are almost always wrong. So, don't dwell on it.

Our *Next Wave* article, *I Can't Believe They Didn't Like It: Part I, Manuscripts*, reproduced in *The Extras Section*, adds a brief overview about responding to manuscript critiques and approaches to getting your paper accepted.

EXERCISING YOUR SCIENTIFIC JUDGMENT

Science is continually monitored by peer review. This occurs when we submit manuscripts, grants, and come up for tenure and promotion. If asked to participate in any of these events, you must be fair and impartial and exclude yourself if you think that you may have a conflict of interest. Discussed below are suggestions and information about being a reviewer for manuscripts and for grants. The opportunity to serve in this capacity is a privilege but it takes time. Sometimes lots of time.

Reviewing Manuscripts?

You have just received a fax from Dr. Reed N. Wright, the editor for the *Top-of-the-Heap Journal*, requesting that you review a manuscript in your field. After you agree to review the work, you may wonder, "What criteria should I use to review this manuscript?" All journals will provide you with a set of guidelines to be followed. These are typically found on their website. Most journals want to know if a manuscript:

- is in the top or bottom third of the field
- represents novel work
- is paradigm shifting
- represents an incremental advance
- is technically sound
- contains data of high quality
- conclusions are supported by the data
- is only technological or descriptive
- is appropriate for this journal or a specialty journal

When reviewing a manuscript it is important to keep an open mind, even if you do not like the title. The most important criterion to consider is whether the data support the conclusions. In making your decision on whether or not to recommend acceptance of a paper, consider the journal and where you place it in the context of all journals. For example, you should consider recommending acceptance of manuscripts for the Top-of-the-Heap journals only if the work is critical to the future of its field or is paradigm shifting. Even if you are excited by the work, it still must be scientifically sound and the appropriate controls should be shown.

While the Top-of-the-Heap journal articles seem to be plentiful, they represent a small proportion of the total science published annually. The majority of articles are published by the journals of the various scientific societies. Articles in these journals are mostly excellent and represent the mainstream scientific achievements of a discipline. Because this is the mainstream of a field, you need to ask whether the work is sound and advances the field. If it does, then you should lean towards recommending that the work be accepted. As the journal that you are asked to review for becomes more and more specialized in its content, the work can be less ground breaking in its content and conclusions. However, the science should still be sound.

Sometimes scientists tend to be overly critical of their competitors' work, especially when they are asked to determine if it should appear in one of the Top-of-the-Heap journals. Consider the fact that if your field never publishes in the best of journals because its members are always rejecting each other's work, then your field never gets the publicity that it may deserve. A consequence of this is that the field is not established enough, and when your lab has the major breakthrough the Top-of-the-Heap journals may not be interested.

Comments to Authors and Editors

Remember it is not your job to rewrite the manuscript or redesign the experiments. If you feel that the manuscript requires a control or additional experiments before you would recommend acceptance, briefly describe the experiments and why you think they should be done. Be concise and very clear. Do not write your comments in an abusive tone. Remember that the authors are going to read your comments and attempt to satisfy your concerns. Like you, they want to publish their work and move on to the next set of experiments and not spend their time repeating their work.

Most journals allow reviewers to make specific comments to the editor that are not seen by the authors. This provides you with the opportunity to provide the editor with a remark that you could not have provided the authors. This section also allows you to justify your view on acceptance or rejection of a paper. While most reviewers spend their time justifying why a paper should be modified or rejected, championing a paper for acceptance is equally important. This is especially true if the work is going to one of the higher-ranked journals.

A detailed discourse on how to review manuscripts, entitled *To Accept or Reject*, is reproduced in *The Extras Section* and was originally published in *Science's Next Wave*. This topic is also mentioned again in *Chapter 9* in the discussion of national service.

Reviewing Grants

"Hi Dr. Starr, this is Dr. Moe Munny from the NIH. I run the Signaling study section at the NIH and you have been recommended to me as a potential reviewer. Can you help us out?" You quickly agree because you read *Chapter 3* and you realize that this is a great opportunity to get some experience on the grant process. When the grants arrive, you may stare at the box and wonder, "now how am I going to get through this?" The good news is that everyone who receives the big box wonders the same thing. As discussed in *Chapter 9*, you will need to give yourself plenty of time to complete the task. Count on two full days per application. Because the pile of grants will represent a general field and may not represent your exact specialty, it is best to try to review those that you think will be the easiest first. This allows you to complete some of the work quickly. Read the titles and abstracts of all the grants. This will give you a sense of your task and how much time you will have to spend on each one. The further a grant is from your specialty, the longer the time it will take to read and write the review.

What to Consider

If at all possible, read the grant through without interruption. Doing so will give you the best feeling for the entire concept and all the information and nuances of the proposal. Some people like to highlight sections and/or make notes on the grant so that they can refer back to those sections when they write their review. Hopefully, there will be summary paragraphs that you can highlight and paraphrase in your review. In reviewing a grant pay attention to the following key points:

- Is the system/field/area important?
- Will the questions proposed address important issues?
- What is the significance of the work?
- Is the work hypothesis driven?
- If successful, will the work substantially advance the field?
- Is the experimental design sound?
- Are all the aims strong?
- Is the success of the project dependent on a single aim or reagent that is not well characterized?
- Did the applicant include important controls or experiments?
- Is the project innovative?
- Is the applicant qualified to perform the work?
- How does this grant rank next to the others in your group of grants?

In time, you will compile your own list of key points, and you will develop a feeling for what you look for in a grant and what you think makes a great grant. Most granting agencies will provide you with a format to use in writing your review. Follow it. In critiquing the aims of the application, state the goals and the strengths and weaknesses. Try to avoid describing the entire aim. If you do find weaknesses and think that you have solutions for them, express these points in a clear and concise manner so that the applicant can address the points in her response. Remember that the applicant will address your points in her revision, hoping to satisfy all of your concerns and get a better score if necessary.

After you have written your review, indicate your relative level of enthusiasm. At the end of reviewing all of your grants, go back and read all of your reviews. Were you consistent throughout the process or were you too tough on the first application? Now is the time to adjust your rankings and perhaps your level of enthusiasm. Your study section colleagues expect that you will have ranked your grants and that not all of your assigned grants would be rated as "outstanding."

Study Section

While you may think that grant review can be performed by email, the actual meeting of the reviewers serves several purposes. The first is that it forces the reviewers to attempt to rank all of the grants in a consistent manner. Second, it assures that differences of opinions can be expressed and a compromise reached (or not). Third, it allows the program officers from the funding agencies to hear the conversations and relay the information back to the applicant so that they can improve their scores if they did not get funded.

Because only 10–20% or so of grants get funded, you will need to be a strong advocate for the best grants. If you are the lead reviewer, you will be asked to present a summary of the proposal and present its strengths and weakness and why you ranked it where you did. Try not to read your review to the group. It's boring and you will lose the attention of the committee members. Be prepared to summarize the project and significance of the work, what critical issues it addresses, how it advances the field, why it is innovative, and what you really like (or dislike) about the experimental design. If you like the proposal, be sure to focus on its strengths. Your enthusiasm or lack thereof will sway your colleagues to agree with your scoring. If you do get into a fervent debate over the level of enthusiasm with a panel member, argue your points clearly and concisely. Hopefully, your first foray into the process will go smoothly and you will not have to present first at the meeting. Watch, listen, and learn from your fellow members.

Our *Next Wave* article **To Fund or Not to Fund**, reproduced in *The Extras Section*, provides an in-depth discussion of being a grant reviewer.

Scientific Views and the Public

Because Academic Scientists are truly the experts in a scientific field and technology, you may be asked to comment on television or for the newspaper about a new disease outbreak or technical achievement. If "60 Minutes" calls, consult with your lawyer first. If you do get to express your scientific opinion in public, you must remember that you are also representing your institution. Offhanded comments or "shooting from the hip" can lead to consequences that you do not want to face later. Be thoughtful. That's why they are asking you to comment and exercise your scientific judgment. You are now "The Scholar." Congratulations!

CHAPTER 7

BEING A TEACHER

"**D**o we have to know this stuff for the exam, Professor Starr?" This question will be with you always. You will hear it hundreds of times during your career. In fact, when you were a student, you probably asked this question a couple of times yourself. There are many opportunities for academic scientists to become effective teachers because academic teaching takes place at many levels. From the classroom to the bench top to the bedside, most of your time will be spent imparting your wisdom and instructing others. For most scientists, teaching methods were not part of any curriculum that we studied. Many of us received our first and perhaps only teaching experience as a TA standing in front of a group of eager freshmen all looking into a jar of pickled critters. As in many other disciplines, trial by fire may get the job done, but it does not always make you the best at what you are charged with doing. So, what makes a good teacher? Is it the teacher's ability to present the facts in a logical sequence, pique the interest of the class to a frenzy, provide a simplified version of a complex system, or is it simply the ability to provide a solid 50 minutes of entertainment that students can enjoy and remember? Perhaps it is a bit of all of these. Because "teaching" is an academic discipline of its own and volumes of books and journals have been published on this topic, this chapter will provide a very abridged guide to preparing for your role as a college professor and introduce ideas and practices for you to consider in designing your classes. The topics will include:

- Teaching in a classroom setting
- Visual aids
- Evaluating your teaching skills
- Participating in courses
- Developing a national reputation for teaching
- Record keeping

TEACHING IN A CLASSROOM SETTING

> **Survey Says**: A tenured Professor — "When assigned teaching responsibilities, make certain that you 'teach well' and that you meet all your responsibilities to students..."

Getting Ready

Hopefully, when you negotiated your position you arranged to have a time lag before you have to prepare your lectures and teach your classes. If this is the case, you should find out what classes you will be expected to teach as soon as possible. This will achieve two things. First, it will (hopefully) avoid the last minute issue of: "Oh, why don't you also teach Dr. Brillodooz's lectures?" The second advantage is that it will give you time to think about your lectures, and, if another faculty member is giving them this year, you will have an opportunity to sit in on her lectures, observe her teaching style, and find out what she is teaching. This latter point could be especially important if you are a yeast geneticist and your first teaching assignment to second-year medical students is on fungal infections in humans. Attending the lectures will save you enormous time next year when you take on the lecture topic yourself, regardless of your personal experience with fungal infections. Even if your particular assignment is not yet determined, attending some of the lectures in a class that you know you will be teaching in is a good way to judge the depth of the subject being taught, the aptitude and attitude of the students, and the general tenor of the class.

If you have to teach right away (within the first few months of becoming the NKOTB), you should plan your time really, really carefully. You need to budget your time so that you can get your research program going, as well as prepare for your teaching assignment. The amount of preparative time that teaching assignments take depend on what the course is and who the target audience is. Undergraduate classes and labs require the greatest time and effort. This is mostly due to the large number of lecture hours associated with a typical undergraduate teaching assignment. Medical and other health professional classes take serious preparation, but the number of overall classroom hours is typically small (<8 hours/class/faculty/year), and the subject matter is often directly related to the professor's field. Graduate classes can take substantial preparation time, especially if they are at the introductory level. Upper-level graduate classes that use the literature as its guide require the least preparation. Even so, each class requires several hours of preparation time.

Regardless of the type of class, get the syllabus and the book as soon as possible. Try to get the exams from the previous year and take them yourself. Hopefully you will pass. This will give you an idea of what the previous instructor(s) thought was important. Consider the simple fact that it typically takes 10–12 hours to prepare a new lecture. Yes, it does! If you have three 1-hour lectures to give a week, this means at the very *least* you will be spending three full days preparing your lectures for each week of teaching. Yikes!

Preparing Classroom Lectures

With the textbook, syllabus, exams, and hopefully some experience sitting in on Dr. Brillodooz's classes, you are ready to prepare your first set of lectures. To begin assembling your notes and outline, consider answering the following questions. In fact, you may want to consider these points every time you prepare for a class.

> - What are the objectives for the lecture?
> - What is the background of your audience on this topic?
> - What is the big picture?
> - Which examples convey the information best?
> - What am I going to test them on?

Objectives

Faculty are often given a syllabus with a simple phrase or single word as the topic to cover. This doesn't tell you how to fill the hour. Even if you are an expert in an area, the textbook, old exams, and syllabi will show what others thought were the important areas in that field. Check these out, as they can save you some time. After going through these materials, ask yourself, "What do I really want the students to understand?" Organize these points into 3–5 objectives or questions. These topics are now your goals and should be the organizational points of your lecture. Design your lectures around these topics rather than on *all* the items in the textbook. If there are other areas that are in the book that you want the students to at least have heard of, be sure to assign the reading and tell them, "this will be on the test." As you know, if you don't tell them, they will ask.

What Level of Instruction Does This Class Need?

This question forces you to decide how to prepare and the depth at which you will cover topics. Health care professional courses (medical, physician assistant, nursing, etc.), graduate, and undergraduate courses require different sets of facts. The professional courses are designed to take students with baseline knowledge in an area to a point where they are conversant in the field. For example, medical students, physician assistants, and nurses all treat patients with tuberculosis; however, the level of scientific detail that each group needs is different. Most of the students have no substantial knowledge of the microbial and immunological worlds. Thus, the courses must start at the very beginning. The physician needs the greatest level of information, as she will ultimately be responsible for designing the treatment of the patient. The physician will also have to defend the treatment plan on a scientific level to the patient, provide the patient with information about the need for multiple antibiotics, as well as provide him with the prognosis of success of the treatment. Thus, she will need to know something about bacterial load and the rates of spontaneous resistance to each antibiotic.

The physician assistant will most likely be examining the patient and communicating with the nurses and the physician about the patient's symptoms and progress. While the PA requires some level of basic science about antibiotic resistance mechanisms in mycobacteria, it is likely that this information will not be used to its full depth. The PA needs to be able to recognize the symptoms of the disease, how to protect himself from catching the disease, and what the most likely treatments should be recommended. The PA should also know that spontaneous resistance occurs, and because of this process multiple antibiotics should be prescribed.

In teaching graduate students, the appropriate level of instruction is determined by the level of course. Beginning classes in a subject typically cover a wide range of material with the objective of bringing all the students to "the same page" at the end of the semester. To continue the mycobacteria and drug resistance examples from above, it would be appropriate that the mechanisms of drug resistance and how these mechanisms were discovered be included as part of the discussion in graduate courses. Advanced graduate courses are usually based on the most recent literature, and therefore the most recent papers on drug resistance mechanisms should be discussed. For graduate students, conveying the process of how the information was discovered, as well as how to ask the next series of questions, is important for the development of their critical thinking skills.

The Big Picture

Always provide the big picture — the relationship among the leaves, the trees, and the forest. Why? Students need it, and will ask for it. This is a common complaint that students will make in any given class. Saying, "Okay, here's the big picture" will help them realize that the big picture is coming. It is also important to connect the big pictures of one lecture with those of another. If you use slides or overheads, put the words "Big Picture" on them.

Examples

Giving examples provides a way to explore the leaves on the trees in that big forest. You must be careful not to describe each of the leaves on all the trees in the forest. Stay away from providing "all" the details. This is really important when teaching professional health care classes and undergraduates. One in-depth example will suffice. If there are multiple mechanisms and each mechanism is important, one example of each is appropriate. Other examples could be provided through assigned reading if it is necessary that they know what they are. Be sure to relate the example to the tree and the tree to the forest.

The Test

Everyone wants to know what's on the test. They also want to know the format: short answer, multiple choice, or essay. Following the proposed scheme outlined above, the test should focus on the objectives presented. It should test the big picture and the relationship among the forest, the trees, and the leaves. In preparing the exam, be sure that it covers the main points that you addressed in your lectures or assigned as reading.

Exams should be fair, but this doesn't mean that they should be easy. An exam should be designed so that all of the students can pass if they studied the material. Exams should also allow you to distinguish between the top group of students and those that have put in the minimal effort to pass. This can be accomplished by asking questions that require two-step reasoning rather than simple regurgitation of the information. Two-step reasoning questions assume that the student has attained the knowledge base (Step 1) to understand and work their way through the problem that is posed (Step 2) and arrive at the right answer.

Essay style questions provide the greatest challenge for students because they must really know the answer to receive full credit. Essay tests also provide the greatest feedback on how effective you were as an instructor, as you will read in their essays what you taught. You will also be able to view the variation among the students. Unfortunately, you will witness their writing skills and discover the fact that penmanship is an outdated skill. The major disadvantage to essay style questions is the length of time it takes to grade them. Thus, they are best suited to small classes. You will also have to establish a scoring scale for essay questions. One way to do this is to quickly read through all (or many) of the answers without grading them and establish the scoring scale based on what should be a passing or failing grade. Alternatively, you can preestablish a scoring scale by breaking down the question into sub-questions with the points for each part indicated ahead of time. While the latter sounds like a great system, you may find that it will result in lower grades. If you do choose to do the latter, it is helpful to provide the students with the number of points each section of an essay is worth.

Short answer questions also provide you with feedback on the success of your lectures and allow you to test some depth of knowledge in an area. This is often the happy medium between the essay and multiple-choice question formats. Short answer questions require less effort in grading and can be used with a larger class.

An interesting version of the short answer format is the true/false, correct the false format. Here, a statement is provided that is either true or false. If false, the student must make a minimal number of alterations in the statement to make it true. This format is very flexible and it's easy to make up questions.

Multiple-choice (aka, multiple-guess) questions are probably the most common format used. This format allows you to test many topics, and with computerized answer sheets, grading is effortless. You can even use the statistics packages associated with the grading programs to determine if a question was fair or not. Multiple-choice formats also allow you to cover a lot of material. By using the "which of the following is false/true?" type of a question, you can cover 10–15 minutes of lecture material. The drawback to this approach is that there has to be only one correct answer.

> ▪ Students expect that you will write clear and unambiguous questions that test the main points of your lectures. To achieve this goal and to test their knowledge, you will have to spend considerable time and effort.

Styles

A key to enjoyable and successful teaching is to get the students involved. In some large, didactic classroom settings, it is simply not practical to just ask questions of the class, especially if the student body is not prepared. To get the students interested, try relating the topic to their lives. Telling an amusing anecdotal story about the topic will help. If you are funny, they may even laugh. Of course if they don't, sometimes it helps to inform them that your story had a punch line. When I was in college, some faculty purposely dressed funny just to get the student's attention. At least I thought that's what they were doing.

Graduate-Level Classes

As mentioned earlier, the introductory graduate-level courses are typically survey courses that serve to get all of the students onto the same playing field. This does not mean that they are not challenging or that they do not use the primary literature. It simply means that you must start at the beginning of each topic. Saying that the "DNA is composed of four bases ..." would not be out of line on the first day of a genetics class. You can still end up talking about the molecular mechanisms of imprinting and epigenetics later in the semester.

Even in introductory courses, you have a choice between didactic lectures and other methods of teaching. Posing questions to the class within an expository lecture is a sure way to interject some Socratic style into a didactic classroom setting. This will at least keep the students awake. If you choose a more Socratic style, make sure that you assign the reading ahead of time and that the reading assignment is not so onerous that the students simply won't accomplish it in time.

In some introductory classes and in upper-level graduate courses, you can make the students work during class. This can be easily accomplished by assigning papers for the students to present or discuss. In the presentation mode, one or more students present an overview of a paper to the rest of the students. The advantage of this style is that the student presenting learns how to present a paper and critically review it. A downside of the style is that the other students may not have read the paper. If you make the students ask questions, this will start a discussion going, which is what you really want. Another way to generate a discussion about a paper is for you to use the Socratic method of teaching and ask questions yourself. Here are some examples.

- What were the goals of the manuscript?
- Were any special techniques or reagents employed?
- What was the goal of Figure 1, etc.?
- What controls were used?
- What controls were left out?
- What should they have done instead of or in addition to?
- Did these authors do anything else that was noteworthy?
- What paradigm was altered/established such that this paper made it into a Top-of-the-Heap journal?

A great system to decide whom to call on is to draw their names from a hat or empty ice bucket. This serves two purposes. The first is that it identifies the students who have put in the work and who are bright. The second is that this practice adds excitement to the class, as the students never really know when they will be chosen to perform. It can even be amusing when all the students put the name of their buddy with the curly red hair in the bucket.

> - Ultimately, you must find your own voice and comfort level in the way you approach your teaching duties. The important part is to have fun. If you do, they will probably learn something and so will you.

VISUAL AIDS

Chalkboard out, PowerPoint in, right? That all depends. Even though PowerPoint presentations can include animation, the chalkboard still has its place and can allow you to convey complex models and points that other formats cannot. This being said, the ability to incorporate an mpeg of "The Rampage of Killer T cells" or "T cell Massacre, Part III" into your lecture may keep the students glued to the scream, I mean screen. We are now in the world of complex visual aids. You need to use them wisely to communicate your ideas and lessons. Here are some tips for preparing PowerPoint presentations:

- Provide a printed copy of the slides as a handout. Also provide space for the students to take notes.
- Use lots of pictures.
- Many textbooks come with a website or a CD containing their figures. Use the figures in the text when possible so that you, the students, and the text are on the same page, so to speak.
- Reading from a list of points on a slide will turn your audience off, even if you are using PowerPoint and the words fly in from another room.
- Provide your own definitions verbally.
- As with science talks, choose good color combinations. No blue backgrounds with red lettering.
- If you are unfamiliar with the lecture hall and you are using your own laptop, show up early (or even the day before) to make sure that it all works.

How'd Ya Do?

Evaluation of teaching is necessary for your development as a professor. We are all aware of professors who never changed their awful style. They simply have no excuse. The data from teaching evaluations are also sometimes used during tenure and promotion deliberations. So, what is being evaluated? This really depends on who is doing the evaluation. Student surveys akin to the one in **Appendix 7-1** provide the view of the customer. This is the easiest view to obtain. Your colleagues will often use the "student view" to evaluate your teaching proficiency. While you and some of your colleagues may think that the customer doesn't know anything about your subject matter, they are paying in some way, shape, or form for the education that you provide, and therefore their views should be considered. Having said this, you will need to devise your evaluation materials so that they focus on the course and not on your bubbling or grumpy personality. Of course, if the students comment that you are grumpy, unpleasant, and not interested in them, then you should carefully review your approach. Many faculty use a numerical system from 1 to 5 (1 being the best) to help rank the teaching effectiveness. How well did Dr. Starr communicate the objectives of the lecture? Were the exams fair? Etc. ... Scoring 1s and 2s is great. 5s are not so hot.

Self-Review

Even if you scored great, you should review yourself. Questions you should ask yourself include: Did the students do well enough on the exams to indicate that they truly learned the material? If so, was it the majority of students or a handful? Did you cover all the material that you needed to cover? Do certain lectures need to be expanded? Was the course too easy? **Appendix 7-2** provides a worksheet for you to conduct a self-evaluation of the course with a focus on the topics and your performance.

Professional Review

While student evaluations provide you with the consumer's view, it is useful to have a senior colleague evaluate your skills and style. This can happen informally if you are participating in a large course and the course director shows up to one of your lectures. If this doesn't happen, you should invite a senior colleague to evaluate one of your lectures. Ask her if she has suggestions. As always, when asking for advice, don't argue with the critique, but instead seek how to improve your style and presentation.

Some schools have teaching evaluation committees that observe and review the lectures of faculty who are coming up for tenure. If your institution uses this system, try to schedule their observation with your best lecture, as determined by your previous student evaluations.

> ▪ In short, ask yourself, "how can I have fun with this course?" This is one of the most important aspects of the task. Enjoy yourself! This is what academic life is about — the dissemination of knowledge. If the instructional format is boring, change it.

PARTICIPATING IN COURSES

"Hey, Ima, you work on Shaadupps, right? How'd you like to take on my lectures on emerging pathogens in The Bronx? I heard that Shaadupps are real common there." Except for the fact that you don't know where The Bronx is located, how do you respond to Dr. Kneematoad, when his request may in fact be reasonable? To circumvent the possibility of being assigned courses or lectures that you do not want or cannot teach, you will need to know which courses you will enjoy teaching, which ones will help your career, and which ones are necessary to get promoted. Of course, it is best if these three categories overlap.

To get an idea of which courses you will enjoy teaching you will need to find out about the curriculum that is sponsored by your department. If it is an undergraduate department, then (hopefully) the courses that you will participate in were spelled out clearly when you joined the faculty. If they were not spelled out, you should attempt to choose wisely. Look at how the classes are taught with regard to faculty participation. Are classes taught by one professor or by a team of faculty? The difference is not only in the overall workload. Being a member of a team may provide you with a buffer (time) to figure out how to approach your teaching. Following the lead or style of your experienced colleagues may make your first experience easier.

If you are in a department that is in a medical school, then your teaching is likely to be in medical school courses. Most medical school departments are responsible for at least one or two courses a year. The courses are usually team taught, and each faculty member has only a few lectures. In these situations, you may be able to request a particular set of lectures. You may also be able to coordinate your lectures around grant deadlines, etc.

Regardless of your departmental affiliation, graduate school teaching is where you have the greatest opportunity to control the types of classes that you will participate in and the curriculum. Participation in graduate-level courses can also be used to your advantage. For example, by teaching in graduate-level classes that are taken by first-year students, you get to advertise to those new students that you exist. Incoming students are usually very open minded about whom they want to work for. Thus, you now have an opportunity to show the incoming students that you are wonderful, charming, and brilliant.

Advanced courses provide a different benefit to the faculty. Advanced course topics are typically based and designed by the instructors to fill a niche that they feel is important. Such courses use the primary or review-based literature as the basis for the classes and discussion. The advantage to participating in these classes is that this allows you to brush up on an area, stay abreast of a field, or learn a new field. All of these can help your research. The literature *survey* may even help you get some new ideas for that hot Specific Aim you want to include in your next grant application. Some faculty like to coordinate general review courses where they use the "Annual Reviews" series of their favorite field as the curriculum and text for a class. This way the class changes every year and they become an expert in 12 or more topics in a field.

> ▪ If you have to teach a course, choose to teach in an area or a type of course that will benefit your research and your interests.

Our *Next Wave* article **To Teach or Not to Teach?**, reproduced in *The Extras Section*, more fully discusses why you should teach and how to approach your first set of classes.

NATIONAL REPUTATION FOR TEACHING

Some institutions will promote faculty to the tenured ranks or promote from associate to full professor on the basis of excellence in teaching. If this appears to be important for your own promotion and tenure, then it is critical that you know what your institution considers to be the measure of excellence. Not only will excellent teaching as evaluated by your students and colleagues be considered, but some institutions may require that you develop a *national reputation* for teaching.

A national reputation means that you contribute to the education of students, fellows, physicians, and others outside of your institution. There are many ways to establish this reputation. The easiest, of course, is to write a textbook that everyone uses at the graduate, medical, or undergraduate level. Okay, so that's not easy. Below is a list with some descriptions of what can be considered as evidence for national teaching contributions:

- Teaching courses at scientific meetings
- Participating in training courses or workshops at institutions like Woods Hole or Cold Spring Harbor Laboratories
- Participating in writing questions for National Board exams
- Developing new models of instruction in professional courses that are disseminated to other institutions
- Developing or participating in public education programs through the media
- Publishing articles on teaching techniques
- Publishing magazine or newspaper articles for the general public on aspects of science

If being an outstanding teacher is going to be your path to promotion and tenure, be sure to look at what others in your department or school have done with regard to their teaching portfolios (see below), as this will be your guide to staying on track.

KEEPING RECORDS OF YOUR TEACHING

Keeping good records of their teaching experiences is one thing that most research oriented faculty fail to do. When asked to provide a synopsis of the courses and lectures that they have taught, they have trouble remembering, especially when it comes to those lectures given in the early years of their appointments. Like your overall CV, your teaching CV should be continuous. The sections should be broken down into students/fellows trained, thesis committees, lecture courses, medical instruction (if appropriate), and educational seminars or workshops organized and taught. **Appendix 7-3** provides a brief Teaching CV that you can use to track your activities. Start using this form in your first year and keep it up to date by periodically filling in the information.

> **Survey Says**: Tenured Professor — "Teaching is critical for promotion [at this institution] and most faculty members don't really understand what that means ... For example, faculty with a great deal of grant support may do few lectures, but have a lab busting at the seams with graduate students and postdocs. They forget that is teaching and don't document it very well."

You should also keep your evaluations, lecture notes, and syllabi for the courses that you participated in. Organizing them either by year or by course will help you when you need these records for your promotion, as you will read about in *Chapter 10*. Even if your current institution does not require these items, you never know where you may take a different job. It is possible that your next appointment will require some documentation of your great teaching ability.

> - Even if you work at a research-centric institution, teaching can be a very rewarding experience. It is, however, up to you to make it so. Enjoy!

CHAPTER 8

MENTORING
1 – ON – 1

"**G**ee Dr. Starr, can I work in your lab for my doctoral dissertation? I'm really very careful and I am sure I won't break as much equipment as that student with the curly red hair did last year." Before you jump and say SURE, you should consider that training a student is a serious time commitment from you. One-on-one refers to the most advanced level of all education: the training and mentoring of graduate students and postdoctoral fellows. Your expertise, knowledge, and ability to encourage and guide are conveyed directly to the student. You are the Master. This type of training occurs primarily with graduate students and postdoctoral fellows, but can also occur with undergraduate students and medical students and residents. Each of these types of students has distinct goals. It is important to recognize these goals and help the student achieve them. Sections in this chapter will include:

- Training graduate students
- Training postdoctoral fellows
- Training undergraduate students
- Training students how to write papers
- Teaching presentation skills
- Providing advice

GOALS

Graduate students and postdoctoral fellows have distinct goals and expectations. The graduate student's goal is simply to earn a doctorate. Of course, to do so he must learn how to ask scientific questions and how to solve those questions. Postdoctoral fellows expect to become an expert in an area and a discipline, to carve out their own identities, and ultimately to be competitive for whatever jobs they are interested in pursuing. Thus, while some aspects of how you teach each of these individuals is the same, the relative degree of freedom and responsibility that each has is different.

GRADUATE STUDENTS

The relationship between a graduate student and a thesis advisor is symbiotic. Graduate students provide "hands" to perform experimental protocols, but they also provide young, creative minds that add to the development of a research program. Whether a student performs well or poorly, he/she will have a tremendous impact on your program. The key is to figure out how to turn the graduate student into a functioning junior scientist. To do this takes time and effort on your part. It also requires some strategy and planning.

How much time does it take to train a student? If you estimate the number of hours that you will spend with a student designing experiments, interpreting data, writing manuscripts, and of course imparting your wisdom of life and science, the weekly hours will add up quickly. For example, if you spend 4 hours a week with a student you will use 208 hours of your time each year. A student taking 5 years to graduate will use 1040 hours or 6 months of your working life. You may find that this number is on the low side, especially at the beginning of your career. The reason for adding up these numbers is to provide you with a warning. If you fill your lab with lots of students, you may find that: (1) you can no longer perform your own experiments because you are always in your office discussing experiments with your cadre of students; and (2) while your students are developing their research skills, your research program may go from a nice steady trot to a slow crawl or even to a standstill.

> ■ The advice here is to start off slowly and do not take on more students than you can handle.

> **Survey Says:** An interviewed tenured Associate Professor stated in response to a question about the number of students that a junior faculty member should start with ... "take the number you think you can supervise and divide it by two."

With the above warning considered, you must also decide how you are going to assign a project to a student, coordinate his research, and be sure that he finishes in a reasonable time frame (5–6 years). As discussed in *Chapter 4*, there are many ways and styles in which faculty work with their students. Regardless of your style, there are six areas that contribute to graduate training.

> - Choosing and designing dissertation projects
> - Technical instruction
> - Evaluation and interpretation of data
> - Reviewing the literature
> - Writing papers
> - Presenting results

Dissertation Projects

So, how do you assign a project to a graduate student? Ideally, dissertations have a common theme or thread, and have as its basis a topic of interest that is pursued through long-, medium-, and short-range goals. However, in some cases the thread is not apparent until the thesis is complete, and may require considerable ingenuity on your part to spin together. Using your RO1 as a guide can help you initially formulate each of these targets. The long-term goal may be an Aim or two on your RO1. This provides the student with the "end-of-the-road" view and provides him with a guide to describe what his dissertation is about. Medium-range goals can be thought of as a series of experiments that lead to a publication. The number of questions or hypotheses to be tested in these medium-range goals should be small and take from 6 to 18 months to complete.

Short-term goals should have weekly or monthly timelines. They should have clear endpoints, with the opportunity for review and evaluation. As the student gets better at achieving his short-term goals, then his project can be expanded, allowing him to exercise his independent creativity.

In the beginning, graduate students require precise details in designing and performing their experiments. The tasks should be organized so that several days' work is planned out, with written protocols provided. Additionally, you should make sure that the students understand why they are doing each step. This will not only help their development by increasing their technical knowledge, but will allow the students to begin to participate in the science.

Proof of Principle

One of the most important aspects of establishing a plan leading to a student's dissertation is to determine whether the project has a sound basis. While some projects may be direct derivatives of your work, others will travel down new roads. If your student is going to embark on a new road, you need to make sure that the experimental design, plan, or goal is fundamentally sound as soon as possible. Proof of principle must be established early. For example, examining the regulation of a gene at the transcriptional level is only a realistic goal if the gene is (1) regulated, and (2) regulated at the transcriptional level. If this point cannot be established, then this project will ultimately fail and you and the student will be punting (See *Chapter 4*).

The Fishing Trip

It is often tempting to have a new student spend his first year or so exploring new avenues of research. A common example of this is to have the student perform a screen for new genes, mutants, antibodies, hybridomas, interacting proteins, etc. Each of these so-called "fishing trips" is high risk, but if successful will lead to new discoveries that may make it into the next issue of *Top-of-the-Heap Journal* and a wonderful dissertation. There are two potential problems associated with sending students on fishing trips. The first is that these projects are high risk, and if unsuccessful the student will be no further along toward his dissertation work then when he started. This being said, the student will have learned a lot about a set of technologies and will have increased his overall skill level. While this is important for his development, it may unnecessarily delay his graduation by a year or so.

The second problem relates to the loss of people-power years in projects that dead end. Because most tenure-track faculty are on a short time clock, the loss of people-power years can be critical to the success of the junior faculty member. Thus, if your work requires such screens, then you need to establish clear timelines for progress, reevaluation, and redirection. Said in a different way, you need to establish a specific time frame for fishing, and be ready to cut the line and move on if the fish aren't biting.

Multitasking

Productivity is increased by the ability to overlap experimental procedures during the course of a day. Most new students (and some older ones) have trouble organizing their time in such a way that they can be productive. For example, if you find your student sitting at his desk and you ask him what he is doing, and he says, "I'm digesting," referring to the fact that he has a restriction enzyme reaction incubating, you now know that he is not a multitasker. You may even find that he has nothing else planned for the entire day, because after the digestion is complete, he will run a gel, which he has yet to pour. While he will hopefully derive some data from these experiments, he wasn't very productive and could have accomplished a lot more during this day. Introducing students to the concept of multitasking early will ultimately lead to very productive people. One way to introduce students to multitasking is to have them keep a weekly timeline. This way, they can write down the anticipated time for an experiment and have other procedures starting and stopping along the way. A large monthly desk calendar can help serve this purpose. However, you must be careful not to overload their multitasking processor to the point that it crashes and the reboot means losing the entire month.

Multitasking refers to assigning students multiple projects. Providing students with multiple projects (i.e., projects with distinct goals) is one way to have a safety net for both you and them. If one project doesn't pan out, hopefully, the other will. It is ironic that in many cases the second project tends to be the one that provides the bulk of the data for student theses. The only caveat to this is that the student must have demonstrated some level of technical proficiency; otherwise, both projects will fail. In beginning the second project, make sure that the design/protocol does not take priority over the primary project. Second projects are a good way to provide a fall-back on those fishing trips, too.

See our *Next Wave* essay, **The Red Herring** (reproduced in *The Extras Section*), for a more in-depth discussion of student's projects and what to do if they go awry.

Thesis Committees

Everyone seems to avoid student thesis committees meetings. Students avoid them for two reasons: they are nervous about being reviewed, and they know that they have to have their science organized and formatted for presentation (see below). As a mentor, thesis committee meetings provide a safety valve for ensuring that the student is on the right track. Thesis committee discussions are extraordinarily valuable for helping with the direction of a project, sorting through problems, and critiquing the work before manuscript reviewers get their chance. Thesis committees can also help provide the motivation for a lazy student. This being said, you should make sure that your students have regular meetings with their committees — at least once a year.

> ■ Use the student thesis committee forum to review your student's progress, to develop new ideas, and to critique the data that your lab is generating.

Technical Instruction

See one, do one, show one (aka, monkey see, monkey do, monkey show another monkey) is by far the best way to pass down technology to members of your lab group. Performing a protocol with the student at your heels allows the student to see everything that you do. It will also impress the student when your experimental data are so squeaky clean and pretty.

There are two tips here. The first is to have all the solutions/reagents/equipment that you need ready to roll. This allows fluidity throughout the protocol, prevents distractions from the procedure, and ensures that you do not cut corners. The second tip is to have the protocol written out ahead of time so that the student can make notes on the protocol itself. This way your student can catch the nuances of your technique and write down what you say as you do the work. If your lab is large enough you probably are already using the see one/do one/teach one strategy to spread techniques through the lab. This works, of course, as long as protocol drift is kept in check (see *Chapter 4* and *Lexicon*). It is important to note that, if the student cannot perform the protocol to your satisfaction on his own, you may have to be the one doing the "see one" part so that you can catch the errors.

If you are developing new protocols in your lab, then you have several choices on how to proceed. Some students, techs, and postdocs really like the challenge of creating the protocol from the literature or traveling to the other end of the hall (or country) to learn a new technology. If your student has the talent to develop a procedure, let him go for it. To help him along the way, discuss the methodology in detail with him, and try to restrict his first series of experiments to include only those that demonstrate that the procedure is reproducible and accurate. Remind him that he should have his reagents ready to go before he actually begins the protocol. If he is having problems getting the protocol to work, you should consider asking a faculty member who has the protocol working in her lab to meet with you and your student to discuss the protocol and your results. Most faculty are more than happy to spend the time.

If your student cannot get the protocol working or if the technology is sufficiently complex to prevent a quick start-up, you may consider sending the student to a lab outside your institution to learn the protocol. This practice is common and can result in a wonderful collaboration. There are two added bonuses here. The first is that the work will move forward very quickly and your student will be able to bring the technology back to your lab. The second bonus is that your student will see how students in other schools work and will more than likely come back energized and ready to crank out data.

Evaluating and Interpreting Data with Students

There are many aspects of evaluating, interpreting, and critiquing data. This is what we do as Academic Scientists. In evaluating the quality of data with students, it is important to communicate with them why you think the data are great or lousy. If the quality of the data is not perfect, it is important to lead them through your rationale for changes in the experimental approach.

Students like to try to interpret their data by themselves. Most of us are guilty of viewing the data as they are produced and exclaiming our interpretations before the student has a chance to breathe. While this is exciting, it is bad for student development, in that they need to think about their results, make their own conclusions, and decide if their controls are valid. (My students told me to write this paragraph.) Of course, if they haven't figured it out in 15 seconds, then go ahead and jump in.

Learning Through the Literature

Bringing your students up to speed and keeping them abreast of your and their field is an important part of their development. Because young labs have inexperienced personnel, they suffer from not having the scientific background in an area to make contributions to future experimental directions and to the interpretation of the data. To overcome this deficiency, you can assign papers to your students for presentation to your group. The presentations do not have to be extensive, but only long enough to highlight the important points and the technology that is being used. After a few months of this, you will find that your lab members will have a better understanding of where they are in relation to others in their field.

A great way to improve on this point is to take students to scientific meetings. This serves four purposes. The first is that the students will be exposed to the whole field all at once, which will help to bring them up to speed in under a week. The second is that the students will meet the competition. They will find out that the competition works harder than they do and is determined to scoop them really, really soon. Hopefully, this will motivate your students to work harder. Third, your students will begin to place the faces of scientists with the names in the journals, which will make what they read real and important. Finally, they will also find that science is exciting and fun.

POSTDOCTORAL FELLOWS

Postdoctoral fellows are key to the scientific growth and success of laboratories. Postdocs not only provide experienced hands but they should be capable of independent research, establishing new technologies, and, of course, producing oodles of data for both of you to think about. If you, the junior faculty member, are lucky enough to attract a fellow, you should have a training plan in mind.

The first aspect of a postdoctoral training plan is to have your new postdoc apply for her own money (if she is eligible). The reason for this is that it will force her to read the background literature and to create a research project that focuses on your general laboratory goals and aims. Essentially, it forces the fellow to write a research plan for her first year or so in the lab, thereby establishing short, medium, and long-term goals. If she is lucky enough to have the application funded, then you now have extra money to spend on additional personnel or equipment.

If the fellow is ineligible for extramural funds, then a plan of attack that establishes short-, medium-, and long-term goals should be formulated. As noted in the section on graduate students, using an Aim of a grant application as a starting place is a good idea. Unlike the graduate student, the postdoctoral fellow's plan should not require as much detail on your part and should allow her to use her expertise in planning the specifics of the experimental design. The proof of principle and fishing expedition remarks from above still apply. However, one important difference with regard to fishing with postdocs is that they are technically experienced and have a better chance of catching fish than do new students. Again, if you choose a fishing trip, be sure you have other projects ready to go or fall back on.

After establishing themselves in your lab, postdocs should be able to create a second project from their work. They may also want to branch out and begin to carve out their own niches in the field. This can serve both of you well.

One aspect of postdoctoral training that differs from graduate student training is that postdocs don't have the benefit of the review that occurs with thesis committees. To help overcome this your postdocs should participate in the research/work-in-progress meetings that are usually associated with the graduate programs. If your program does not provide an opportunity for postdoc presentations, then you may want to suggest it. Otherwise, if you think that a group advisory meeting would benefit the progress of the postdoc's project, ask some of your colleagues to meet for an hour to discuss the data, which the postdoc can present. Another idea is to arrange with other investigators who share your research interests to establish joint group or special interest meetings. Such meetings are beneficial to all and will provide a forum for your postdocs to present their work and get feedback.

> ▪ The key to postdoctoral training is to provide a productive and exciting environment in which postdocs can flourish and can establish and meld their own independent ideas with your research goals.

UNDERGRADUATES

Many institutions offer undergraduates the opportunity to perform research as part of an honors program. Additionally, undergraduate summer research programs are orchestrated by schools as a means of providing research experiences to individuals who attend schools without a research base and to attract potential graduate students to the institution. The motivations and goals of undergraduates seeking research experience are diverse. Some may wish to go to graduate school and may realize that to be considered for admission to a top graduate program they need to have a meaningful research experience. Some may want to enter medical school. Research-oriented medical schools like to have applicants who have performed research projects, and this may give a student an edge to admission. Some undergraduates are unsure about what career path they should take and feel that a research experience will help them decide. Whatever you may feel about these different motivations, the goals of the students will be similar: they want to learn something about real scientific research.

Choosing an Undergraduate Research Project

The choice of a project differs depending on the length of time that the student has available to work on the project. A summer student will only be able to carry out one or two short-term goals. The project should have progress points along the way. An example would be to have a student clone a gene into an expression vector of some kind, express the gene, and determine the consequences of the expression. Such a project has three steps, each of which could take the whole summer, but if everything works out well the student may even complete a figure for your next paper.

If the student has a year in the lab, the above project can be expanded to include additional steps. As the student makes progress, you can add to the project. Some undergraduates can spend lots of time in the lab and by the end of the year, could be proficient in a number of techniques and could contribute to your program. Other undergraduates have only a limited amount of time to spend in the lab. It is important to recognize this early on, otherwise, you will be disappointed if they can only spend 10–15 hours/week in your lab. If the student cannot put in more than 15 hours/week, then he should not be involved in working on something that is critical to the success of your research program.

Students that have two years to spend in the lab can be quite productive in their second year, and you will be able to develop a more sophisticated project with them. This of course is the best situation and is likely to have the best reward to you and the student.

Working with Undergraduates

For the most part, undergraduates should receive training that is similar to a novice beginning graduate student. The student should have clear technical instruction, a full description of the research project, training on interpreting their data, and some aspect of training them how to write and present the information. Many labs couple an undergraduate with a graduate student or a postdoc. This works out well, of course, if (1) the trainer is a willing participant and (2) the trainer is technically capable.

It is important that the student understand why the project is being carried out and what the different steps along the way mean. While it seems like these points are easily conveyed to the student, they are not. A student's understanding of these points is what separates the really good students from the pack during their interviews for graduate school and MD/PhD programs.

TEACHING YOUR TRAINEES TO WRITE

When did you get your first real writing experience? If the answer is when you were writing your first grant application, then you already know that that was too late. Students and postdocs need to learn how to write. *Chapter 6* has a section on strategies for you to use on writing papers with your students/fellows. The strategies discussed in *Chapter 6* focus on when to submit a paper, who is going to write the paper, organizing the manuscript, and where to send the completed version. This section will focus on strategies to improve the writing of your students and fellows.

The key lies in first determining the strengths and weaknesses of the student/fellow's skills. Student writing strengths could include: the ability to organize a scientific paper; knowing what to say, how to say it, and when to say it; having the literature base knowledge to place the work in the context of the field; and keen grammatical skills. Weaknesses in their writing skills would therefore be just the opposite of the above statements: lack of organization or knowing where to begin; not having a clue about what to say or how to go about describing the work; a poor foundation in their field; poor grammar; and sloppiness. Fortunately, there are remedies for all of the above weaknesses. The following paragraphs/sections describe some of the common problems and suggestions on how to improve your students' skills and perhaps your own.

Where to Begin and How to Organize

Writing a paper is a daunting task for a student. Because students read papers from the beginning to the end (at least we hope they do), they initially believe that this is how all papers are written. Thus, starting with a Title, writing the Abstract, followed by a descriptive Introduction and Results section seems like an impossible task to the virgin paper writer. To help them and you organize this task, go through *Chapter 6* and the Manuscript Outline in **Appendix 6-1** with the student. Explain that it is easier to write the paper if the Results section is completed first. This, of course, requires that the figures be in near-complete form. If the student is still gun shy at this point, suggest that he outline and write the Materials and Methods first. Because this latter section describes what they have been doing for the last year or so, students are most comfortable writing this section. If this approach is not working, then you may try diagramming a manuscript that you wrote with your student. Just as you diagrammed a sentence in high school, the objectives of each section of a paper can be marked and explained with the manuscript outline form. After this is complete, have the student diagram a manuscript from the literature. With the paragraphs broken down, the student should be able to move on to the next step, creating the outline.

What to Say and How to Say It

Assuming that the outline described above is done, the next step is obviously to take the outline and transform it into full sentences. If the student is having trouble at this step, he should begin by listing all the points that he wants to convey at each part of the outline. At some point, he will have all the issues stated. He can then convert these points to full sentences and try to link them together.

To address technical writing issues, such as how to write the Results in past tense, etc., have your student consult "scientific" technical writing books, a number of which can be found. Information can also be found with the Instructions to the Authors provided by all journals. Additionally, some of the larger scientific societies have programs, books, and information on their websites to help you and your students with the task. There are a number of books, including *How to Write and Publish a Scientific Paper* by Robert Day, that will help you and your students avoid tension about "tenses." Some societies offer courses on writing at their annual meetings. If one of your students has problems writing, then sending him to one of these courses before a meeting would be a great investment in both his education and in his productivity.

If your attempts to guide the student through the process are not working, you should write the paper yourself and have the student follow along at each draft version. Allow the student to comment on the organization, interpretation, and grammar so that he will get a sense for the process. Hopefully, he will do better on his next paper.

Grammar, Style, and Sloppiness

While some try to combine these three elements into one, they are distinct. There are specific grammar rules. If you believe that your student has poor grammar, and yours is excellent, then you may be able to teach him the differences between which and that, because and since, or how to avoid the infamous split infinitive. If your skills are not that great, then you should rely on a book on grammar. Several small pocket books are available that can provide you with the rules. A recent book — *Eats, Shoots & Leaves: The Zero Tolerance Approach to Punctuation*, by Lynne Truss — is a wonderful read and provides a terrific review of the rules. Many computer word-processing programs have grammar tools, which are useful too. Such tools may have settings that enable technical writing styles over prose that should be selected for papers and grants.

Style issues sometimes create arguments between students and their mentors. Most of the time, these points are trivial and the stubbornness of each individual is quickly revealed. The only objective should be that the paper be grammatically correct when submitted. The reason for this is that each journal uses a variety of copy-editors who each have their own styles or who have interpreted the style guidelines of the journal in their own way. Thus, after you have argued with your fellow until both of you were unhappy, the copy-editor may choose a completely different style.

Sloppiness is the easiest issue to deal with. If a student or fellow is sloppy, make him put in all the corrections and read the manuscript backwards.

Good luck, and may the Force be with you.

TEACHING PRESENTATION SKILLS

Students and fellows need to get experience in presenting their data in oral formats. For graduate students, this is usually accomplished by their presenting their work at the research-in-progress meetings. To help guide them through this process, many faculty spend time listening to their students practice their talks — sometimes over and over and over again. While it is *very* tempting to interrupt your student during these practice sessions, it is more useful to jot down notes along the way and then discuss the corrections/suggestions at the end. Beginning students may require that most of the information that they are going to present is outlined on the slide. While this may initially be boring to the audience, it helps the student organize his thoughts and makes him more comfortable. As stated above, postdocs have fewer opportunities to present their work. If the opportunities exist at your institution, you should encourage your fellows to participate in venues where they present their work.

Some scientific meetings, like FASEB, Neuroscience, AAI, ASM, and others are designed around small group breakout sessions in which students and fellows can give talks. This is a golden opportunity for your fellows and students to get exposure and practice giving presentations. As above, spending time with your students on their presentations is critical to their success, and ultimately to showing off your laboratory's achievements.

FUTURE PLANS

What type of advice do you give a student/postdoc who is leaving your lab after completing the program? This really depends on what the student wants to do next and where they are in their chosen career path. If the advisee is an undergraduate and the student wishes to study a specific aspect of science, then you can help her locate the appropriate institutions that have programs and will train her to the best of her abilities.

Before you can give advice to a graduate student who wishes to continue his training as a postdoctoral fellow, he must decide on a specific field. Often students will have very specific ideas in mind. Other times they may just have a broader sense. In some cases, students have to (or want to) move to a certain area of the country, restricting their choices. Regardless of how the students generate their list of potential postdoctoral training environments, you will have to help them choose.

One way to guide them is to break down the potential laboratories into the following categories: quality of their average publication (high- or low-impact journals), level of funding, size of the group, publications/year, and special technologies. Organizing the laboratories into these categories allows the students to decide what type of environment they would like to be in. Biggest and richest is not always the best. Some students do not want to enter the cutthroat worlds of the super-large labs, where they will be totally on their own for guidance and advice. Other students can't wait. It is important to remember: it is the student's decision and life. So, give your advice and let the student decide.

"Not all trainees will grow up to follow in your footsteps." Statements like the above are common these days, as the number of faculty positions that are available do not match the number of postdocs seeking positions. With this knowledge in mind, only a small number of your students will ultimately have a job like yours. Therefore, if a student has an inclination toward another area in which he can use his scientific training, he should be encouraged to find out all the information that he can about that area. If you have friends that are in that area, see if your friend can advise the student on how to accomplish his goals.

One of the biggest concerns of mentors with postdoctoral fellows is how to help them get a job when they complete their training in your lab. If the fellow is seeking an Academic Scientist position and has accomplished enough to earn one, then the sticky issue of "what she can take with her" will come up. A rule of thumb should be that if the fellow worked on an Aim in your grant and did not carve out a new niche to follow, then she either will compete with you on the subject at hand or should switch to a related area.

It would be unfair of her to ask to take your ideas and continue to develop them on her own, unless you are no longer interested in that pursuit. On the other hand, if the fellow took your initial idea, went off on a tangent, and did in fact carve out her own niche, then in all fairness, she deserves the right to explore her undiscovered territory. Your investment in her is paid off by the publications and fame that she awarded you as the two of you pushed that new area in science. Often the fellow and faculty member come to an agreement on how to pursue the next set of aims. This being said, many of the agreements last only a short period. Hopefully, enough time will elapse for both groups to find their own direction.

> ▪ For the most part, training students and fellows will be a rewarding experience, and when they leave your nest, you will be sad to see them go. If you were a good mentor, your trainees will seek your advice for years to come.

CHAPTER 9

ACADEMIC
SERVICE

"**Y**ou want me to chair the new poster design committee for graduate student recruiting? Wow, Dr. Kneematoad, that's great! When can I start?" Along with dozens of other requests like this one, professors are asked to perform a variety of services. These can range from departmental or institutional duties to participation on national committees and review panels. Although service-related activities are often performed gratis and are typically not the sole basis for promotion, service participation is necessary for the success of institutions and for science. Not surprisingly, the key to your ability to participate in these activities involves management skills. This is both in terms of your time commitment to the service and how you approach your participation and interactions with others. In this chapter, several types of service will be discussed with the objective of providing you with ways to be an effective and efficient participant. The topics include:

> - Department duties
> - Student advisement
> - University committees
> - National service
> - Keeping records

DEPARTMENTAL DUTIES

Even if you are a junior faculty member, you may have to spend considerable effort in the operations of your department; such as administrating an academic program, running a core facility, or performing clinical duties. In each of these scenarios, time management is your key to balance your research and service.

Administering an Academic Program

Depending on the type of program (graduate, undergraduate research, admissions, departmental seminar series, etc.), directing or administering a program adsorbs a lot of your time and effort. To minimize your time and effort, you need to break down your duties into functional units and administer each separately. To successfully administer your program, you must establish a timeline of the program's activities, delegate some responsibility, and represent your constituents. The discussion below focuses on these three issues.

Timeline

Fortunately, most academic programs do not require that all components begin and end on the same dates. Establishing a working timeline of the events and operational components of a program is therefore the first point of business. Using the academic calendar of your institution, enter the important events on the calendar. Estimate the lead time that you think you will need to organize and administer an event. For example, graduate student admissions is a complex process that begins in September and runs through April. During this time period, you will have to advertise your program and distribute applications, collect and evaluate applications, interview students, select students for admissions, and convince them to come to your school. In choosing dates for events, consider the NIH grant deadline dates, which may prevent some of your esteemed colleagues from participating in interviewing and recruiting. Importantly, consider working around dates that affect your ability to function as a researcher.

Delegate, Delegate, Delegate

While the person who convinced you to take the administrative position was delegating authority and responsibility to you, if your task is multifaceted, you too must delegate. You will first need to decide what specific jobs you want to do yourself and what you would like to convince someone else to do; that is, you may need to acquire some of Dr. Kneematoad's behavioral characteristics. Delegate what you can to people you are comfortable working with. Try to find those individuals who are well organized. When delegating a responsibility, you need to provide your colleague with the objectives of your plan and what you specifically want them to accomplish. Does this sound like a grant proposal or the introduction to your lectures? It should. Efficient organization requires good communication. Importantly, once you have delegated a responsibility to someone, let him do his job! Do not micromanage his activities, it will only annoy him and reduce his productivity.

However, if you find that your colleague isn't getting the job done, don't let her off the hook by assuming the chore. Talk to her, find out if she needs help and if she does find an additional faculty member to help her. In assigning jobs, it is critical to provide a due date. Make it a day or two earlier than it really needs to be done, so that when your colleague is a day late your program is still on time. And they will be late.

Representing Your Constituents

Whom do you represent? Depending on your leadership role, you may be responsible for representing a group of people (faculty) to your institution's administration or representing your institution to a group of people (students). In doing so, you will need to place your thoughts and statements in the context of the group whom you are representing. A common example is the dual role played by a graduate program director (PD). On the one hand, a PD must represent the faculty's expectations to the graduate students. That is, the PD must require that the students achieve a certain level of performance in order to earn their degrees. On the other hand, you are also the sounding board of the students who may want to change some hideous (at least to them) part of their program. In this case, it is your job to listen carefully, consider their opinions, and try to improve the program from their point of view. To be successful at managing two such groups, you must be ready to switch hats and your mindset. However, you must always keep the ultimate goal of your position in mind. In this example, the goal should be to provide the best possible education to the students, and provide it in a manner in which they can succeed.

Managing a Core Facility

Core facilities or service laboratories provide a service to their department/institution. These programs are expected to run smoothly. The advice for the job of managing such a facility is straightforward.

- If you are new to running a facility, start small.
- Initially start with those services that can be routinely performed and do not require substantial troubleshooting or monitoring of the personnel.
- Add services only when your current services are trouble free. This will ensure that you can set up the new services without worrying about your "bread 'n butter" service collapsing.
- Do not provide work for free. If your service charges fees to the investigators, charge them. Not charging or providing a friendly discount will lead to your kindness being abused and you overtaxed.
- Be wary of accepting work in the form of "collaboration." If you are truly providing scientific collaboration then this is OK. If your collaboration is in lieu of fees, then it is not collaboration. Stick to the above guide.
- In hiring a technician/manager for your facility, choose the person that has the best organization skills, so that you only have to supervise and help bring new services on-line.
- If your personnel are not performing well, then you will have to let them go and hire new employees (See *Chapter 3*). It is important to remember that core facilities are businesses, not academic labs. Treating it as a business will ensure that it is successful.
- Remember to make sure that you have time for your own research!

Clinical Duties

Practicing physician scientists who have a laboratory or manage large clinical trials represent a group of Academic Scientists who have to divide their time and effort such that both medical and scientific efforts are successful. To provide you with information on how to do it, several physician scientists with different clinical practices were interviewed. The focus of these interviews was to find out how they manage their professional lives. Their advice is summarized briefly in the following paragraphs.

Because you are likely to have participated in a residency/research program, you already know how and where you can squeeze time out of the day for science. You also should already know where your time can be wasted or lost. Thus, the advice is to make sure that your "protected" time — i.e., time devoted to research — is clearly defined. Your chair, your colleagues or partners, and you should have an established calendar of when you are on call and when you are in the lab. This is really critical and for the most part should be established before you take your position.

> **Interview says**: An Assistant Professor who is a physician scientist commented that solid support at home and a loyal secretary are vitally important for guarding her time.

After establishing your clinical time, there are several problems that you can prepare for that may make your ability to spend time in the lab easier. The first of these problems is that sometimes your patients need you when you are in the lab. How can you avoid seeing them when they expect you to show up? One solution to this problem is to expose your patients to your partners or colleagues on a regular basis so that if you can't come to treat them because you are in your "protected time zone" they will be comfortable with your colleague. The flip side to this is that you will have to see your colleague's patients as well.

How many referrals should you take? Expanding your medical practice may be an important goal for you. The trap is that, if you expand too much, your scientific career may suffer. You will have to make a decision on how much additional clinical service you want to perform and how successful a scientist you want to be. One way to keep the size of your practice in hand is to limit the number of referrals that you accept. That is, you should see patients in your specialty and avoid expanding your specialty into other areas that you may be able to handle but would use up valuable research time. You also need to remember that if you consistently refuse referrals, you may not get them later when you need them, and recruiting patients may be difficult. The key here is to have a plan for which referrals you will take and which you won't.

> **Interview says**: A Full Professor, physician-scientist says: "Be better than a generalist. Be realistic and restrict your tasks. Decide early on what you will do and stick to it."

While all of the recommendations above are aimed at reducing your clinical time while increasing your science time, there are two critical points to consider. The first is that if you are going to continue to see patients or evaluate patient samples, you must maintain your clinical skills. So, eliminating all clinical duties is not sound advice. After all, who wants to see a doctor who only practices in October? The second is that because you will be spending considerable time in the lab and not seeing as many patients as your pure clinical colleagues, it will take you longer to establish your reputation as a clinician. If you do your work well and keep up with the latest and greatest procedures and treatments, your reputation will build and you will be satisfied with it.

GIVING GUIDANCE

Undergraduate and graduate students require guidance on an infinite range of issues. The advice may be as simple as which courses to take in their second year but may get more complicated if they need help on choosing a laboratory or a future graduate school. If the advice requested is personal in nature, you the advisor may want to seek advice on how to proceed. In this section, undergraduate and graduate student advisement will be discussed, as well as what to do if personal issues come up.

Undergraduate Advisor

To begin your duties as an undergraduate advisor, you will need to establish office hours. Because your advice is often needed at critical times of the year for the students (i.e., prior to preregistration, registration, and before graduation) you may want to have extended hours at these times and reduced hours at other times. If you have a grant due around one of these times, try to reduce the number of days that you will be available so that interruptions while you are writing will be minimal.

Degree Requirements

If you have never given advice to a student, you will at least need to know what types of advice they will seek. Typically, students want to know what the requirements are for them to complete their degrees and if any of the requirements can be waived. This being the standard fare, it is important that you become aware of what is actually required. Remember that no student likes to come to an advisor who knows less than they do. Course and degree requirements are usually published in a handbook that nobody reads. You may be the first to read it! When you read it, concentrate on making a checklist that you can use with each of your students. There may also be someone in your department who knows all the rules and can be a great source for this information. Find out who that person is and enter his or her email address into your computer. As your flock of students pepper you with questions over the years, you may want to write them down with the correct answers as a "frequently asked questions" document that you can hand to them when they enter your office for the first time (and each subsequent time). They of course will not look at it before they call or stop by the next time.

Advising Premedical Students

Biological sciences majors are likely to be considering a career in medicine, science, or both. This decision influences the types of courses they may need to qualify for admissions. Admission to most medical schools requires a broad range of science courses but usually not the most advanced courses. Many schools have a "Premed" office that can provide you and your advisee with this information. Such offices are typically responsible for compiling the letters of recommendation that are sent out to the large number of schools that students apply to. Students should be encouraged to contact this office early in their studies. The advice for premed students is simple: get good grades (>3.6); do well on your MCATs (>30), try to get some medical related experience; and really know why you want to be a doctor. Medical schools will also require at least three letters of recommendation. To help make contact with faculty on campus, your students should be encouraged to interact with the faculty and to participate in university or community programs from which they can get such letters. Many premed students join laboratories to gain some research experience and of course to procure that precious "letter of recommendation." Such letters are especially useful for medical institutions that have a research focus. It is important to note that, despite the managed-care fiasco and skyrocketing costs of medical education, medical school admissions is still very competitive and a typical school will receive 4–8,000 applications for 100–150 slots each year.

You may be asked, "Dr. Starr, where should I apply to medical school?" Although you may not have gone to medical school or even considered it, the simple response should be "to as many schools as you can afford." Suggest that the student choose a range of schools appropriate for his or her credentials. No matter how strong their credentials appear, premedical students should apply to the schools in the state where they are legal residents. States typically supplement medical schools (even private schools) to admit a higher number of students from their state then from other states. Medical school admissions books and a variety of websites will state demographics for the students to consider. You should also advise your students to apply early, as many medical schools use a rolling admissions process. Those who apply first have the best chance of getting in.

Advising Pre-Graduate School Students

The advice for students considering graduate school is also simple: get good grades (GPA >3.5); do well on your GRE exam (>80th percentile on each part); gain real live research experience; and know why you want to get a PhD. Many programs do not accept students who do not have research experience. Prior research experience is often quoted as the only true predictor of success in graduate school. If your school has a strong undergraduate research program, then suggesting laboratories within your school that match the student's interest should be relatively straightforward with a single caveat.

As a young faculty member, do not under any circumstances tell a student that she should not go to Dr. Kneematoad's lab. Why? Because Dr. K will find out about your comments and be very, very angry. Instead, if you feel that Dr. K is an inappropriate advisor, leaving his name off your list of suggestions will only be a simple error for which you can reply if quizzed, "I forgot." If the student asks about Dr. K's lab, you can simply respond "Dr. K's lab is fine, but is that what you really want to work on? How about Dr. Brillodooz's lab?" Yes, part of advisement is diplomacy.

Students should be encouraged to participate in a research project as soon as they can. It is better for them if they can start on a project in their junior year or earlier, rather than waiting until their senior year. If they wait until then, they really won't have gained enough experience or insight into the research process to help them in their interviews for graduate programs.

If your school's program does not have a strong undergraduate research program, suggest that the student apply to the many federally sponsored summer research programs offered throughout the country. These programs not only can provide research experiences in high-powered labs, but many provide stipends, housing, and organized presentations. A large majority of students applying to graduate programs participate in these programs. Even if your school has a strong research program, the student will gain valuable experience visiting another institution for the summer and participating in a different environment.

Advice for the Pre-MD/PhD

Combined MD/PhD programs offer combined degrees and seek to train physician scientists from the very beginning of their post graduate work. Many of these programs are supported by the NIH and are termed Medical Scientist Training Programs (MSTP). Like other federally sponsored training programs, MSTPs are reviewed every five years by colleagues from other schools with competitive programs. Thus, the funded programs are considered to be the best. Most MD/PhD programs take between 7 and 9 years to complete and are extremely competitive for two reasons. The first reason is that there are not very many positions. Nationally, there are around 900 NIH-funded positions with about 250 entering each year. The second is that all of these positions come with full tuition coverage for medical school and provide a stipend for the student during their studies. Not only do the undergraduates applying have to have good grades and MCATs; it is almost universally required that they have *excellent* research experience and training prior to admittance. If one of your advisees has research experience and can't make up his/her mind about whether to attend medical school or graduate school, the MSTP may be a good compromise. If one of your students is interested in such a program and has limited research experience, you should advise him/her to seek the best research experience available prior to applying.

A very good student will ultimately have several acceptances for MD/PhD programs and may come to you on how to choose. Each of the programs will vary in the organization of the basic science course work, graduate program organization, exposure to clinical experiences when in the graduate years, and on the timing to enter back into the system to complete the clinical training aspects of the combined program. The student should have compiled the answers to these issues on her interviews and, for the most part, you should listen and use what you know of other schools and experiences to advise. Regardless of the organization of the program, the student will be happiest in a program that has some flexibility, and many of them do.

Personal Advice

You may be asked by a student to provide him guidance about a personal issue. You must remember that students have parents and guardians who are responsible for their lives and may have different viewpoints from yours or their child's. You do not want to be in the middle. Thus, depending on how serious the issue is, your best advice may be to recommend that the student seek the appropriate campus counseling service. All schools have some form of counseling service that is available to ALL that need its help.

If you are presented with a student who is complaining about harassment, discrimination, or a similar problem, go directly to your chair about the issue. Do not pass go! Do not try to intervene yourself! It is not your job, and you will become involved in something that you cannot solve and will regret your participation later.

Graduate Advisement

Advising graduate students takes place at many levels, but the key forms of advice are: which courses to take; which lab to go to; where to go for postdoctoral studies, as well as input on experiments and thesis work.

As with undergraduate advisement, graduate programs also have degree requirements. Some of these may be specific to the graduate program, whereas other requirements may be university based. Published handbooks specifically stating the requirements for graduate degree programs will also be available. And like the undergraduate guide, no one will read them. In most graduate programs, there is one faculty member or a small group of faculty who serve as advisors for the newly recruited graduate students. If you are one of these advisors, be sure to read the graduate program handbook. It is likely to state what the core course requirements are and what electives are available for that program. Unlike undergraduate programs, advanced graduate courses are not offered every year, and sometimes, the courses in the handbook are no longer taught. So, it is important to get the current course catalog (or at least know the website) so that when students ask, you can tell them how to find out if a course is offered. If the advisees are your own graduate students, then you want to make sure that they take courses that will benefit their research progress. This may require that you talk with the various course directors or get the current syllabus to find out what is currently taught in the class.

Advice on mentoring and training your own students is discussed in *Chapter 8*.

Writing Letters of Recommendation

One aspect of being an advisor is that someone will want to know what you think of a student that you mentored. In providing your letter or opinion on the phone, first realize that you will seek such opinions in hiring staff for your lab and possibly when admitting students into the programs that your institution offers. Therefore, to be fair to all parties, you will need to rank your students in your mind and be able to determine what their strengths and weaknesses are. With the exception of my students, not every student is the "best in years" or even "outstanding." Some are truly only average. A letter of recommendation should contain the following types of information:

- Describe how you know the student and how you are able to provide your assessment.
- Provide your qualifier of the student's abilities:
  o Best in years, Outstanding, Excellent, Very good, All other adjectives
- Justify the qualifier by:
  o Describing what he/she achieved. In the case of lab work, this should include the project and his/her role
  o Ranking performance with respect to his/her peers
  o Providing a description of the student's unique abilities
- Discuss the student's strengths and, if you are inclined, weaknesses
- Discuss why he/she should be accepted/admitted/hired/awarded (or not) over others
- Describe how well the student gets along with others
- State that you would gladly accept/admit/hire/award him/her for whatever it is that he/she seeks, if you think that you would be happy to do so.

While it is certainly easier to write a glowing letter for an outstanding student, writing a letter for an average or below-average student is difficult and takes time. One approach is to suggest to the below-average student that he seek a letter from someone else, as you think that you may not be the best qualified (or know him well enough) to write the type of letter that he really wants. The other approach is to focus only on his strengths and avoid all references to how poorly he performed. Such a letter is easy to spot. If you feel strongly that the student doesn't deserve consideration for the position that he seeks, you should say so in your letter. Yes, people do write negative letters! And yes, if they are queried on the phone about a student, they provide honest and candid answers, even if they are negative

Thesis Committee Meetings

Thesis committee meetings serve to guide graduate students through their research programs and ultimately judge their work as acceptable for the degrees that they are seeking. While there are some common threads among **all** thesis committees, the interplay between the student, the advisor, and the committee members varies. As a faculty member, you could have one of two roles: the student's direct advisor/major professor/mentor or simply a member of the committee.

As the Mentor

If you are the advisor your role is to make sure that the plan of action is sound. This means that the student research proposal has long- and short-term goals and that there is a series of experiments to show proof of principle as described in earlier chapters. It is often helpful to review the slides with the student ahead of time. This will allow you to hear what the student thinks is the rationale behind the experiments, as this is typically the weakest point of most student presentations. It will also give you a chance to see all the data that the student will present. You and your student should be on the same page prior to the meeting. In other words, the student should know what you think of the project and what your expectations are. You don't want to be surprised when Dr. Kneematoad says, "Well, when do you think you'll will finish up?" and your student answers "in three months." This is of course a problem if you were thinking two years. In summary, your role as the mentor is to make sure that the student is prepared.

There is a tendency in committee meetings for committee members to direct questions to the mentor instead of asking the student. Instead of answering, allow the student to gain experience in fielding questions. If you disagree with the student's response, rephrase the answer in a way that allows the student an escape and shows that you have a greater understanding of the situation. An example of this may be to simply state, "well that may be so, but another interpretation is that ..."

As a mentor, you should also value the committee's comments, as these are likely to be the comments that a journal reviewer may make when reviewing your student's manuscript. Thesis committees provide a free two-hour review of your projects. Take advantage of these sessions.

As a Thesis Committee Member

This is a position of advisor, as well as judge. As an advisor, it is your role to evaluate the data quality and the interpretation. Does the student have proof of principle to continue on this line of study? If not, what key experiment will be convincing? Do the data support the interpretations? Is the system functional? Are the controls solid? Is there a plan with a timeline that will allow the student to complete his degree in 5.5 years? These are the questions that you should be thinking about when the student is making his presentation. It is important to focus on what was done and what the next steps will be. Stopping a discussion to try to figure out why a failed series of experiments, which are now abandoned, didn't work, will waste everyone's time and should be avoided. Focus on the positive and future plans.

Students have the simple goal of completing their dissertations. They will wonder if they have enough to graduate or, if not, what they need. They will be concerned with the length of time they are in the program. These are good concerns, and all students should be encouraged to complete the program as quickly as possible.

In approving a dissertation, you essentially are stating that the student has met the standards and requirements of the institution. What are these requirements? Aside from course work, doctoral dissertations are awarded for the generation of a substantial body of novel work that contributes to the advancement of science. The keywords to this statement are "substantial," "novel," and "advancement of science." How much is enough? When a student does well, everyone is happy and the decision is easy. A student with three first-author papers in leading journals is most likely to have achieved the mark. What if the student has only two papers? What if he has one or none? Unfortunately, only experience can help guide you in this decision. When a student has not had the benefit of outside peer review to judge the novelty of his work (i.e., no first-author papers), then you will have to decide if he has achieved the knowledge base and experience to receive a PhD. Sometimes, students have collected the data and the papers are written and will be submitted. This allows you to judge the quality of the work as presented in the dissertation. Your colleagues on the dissertation committee will have their views. If you are new to this business, it will behoove you to voice your opinion last. You may be able to voice your concerns, but wait on your vote until you hear how your more senior colleagues view the work. Remember to be fair to the student, your institution, and to your own standards. Good luck.

UNIVERSITY COMMITTEES

Universities run through the work of faculty committees. These permanent or ad hoc groups meet, discuss viewpoints, and make recommendations or decisions on a wide variety of topics that range from university governance to tenure decisions to the location of the next departmental retreat. Some of these committees make long-range decisions on the future of the institution that could affect your ability to carry out your research or affect the future of your colleagues. Depending on the committee to which you are assigned, your role, effort, and time commitment should vary with the overall importance of the committee work. Regardless of the committee, there are some rules of advice that should be followed, as they reduce the time spent in the committee and increase the efficiency of both the committee and you. Your ability to enact this advice will depend on your seniority and diplomatic skills. The advice is listed and discussed below. If you are the chair of the committee, you will have more control over each of these suggestions than when you are just a member. Nonetheless, with skill all of these points can be employed successfully regardless of your role on the committee.

> - Start the meeting on time
> - Have an agenda and stick to it
> - Do not tell anecdotal stories
> - Have a plan
> - Have support of others
> - Consider all parties
> - Have an ending time

Being on Time

Always be on time and be ready to start on time. Typically, faculty who do not often get to sit and chat may want to do so prior to getting the meeting underway. While this is fun, it wastes time. A conversation over lunch may be more enjoyable anyway. The advice is to keep your greetings and pleasantries short so that the meeting can convene as scheduled. If a quorum has convened it may be possible to suggest that you start with the real excuse that you have other meetings to attend after this one or that you have an experiment running.

Have an Agenda

Most meetings do in fact have an agenda. If you are asked to participate in one and did not receive an agenda, and you think it will be a long or complex meeting, request an agenda. Requesting an agenda will not be insulting, and the excuse will be that you would like to think about the goals ahead of time. Of course if you request one, you will need to read it and think about the topics. This could also give you time to ask some of your other colleagues what they think about some of the issues.

A problem with committees that deal with the resolution of complex problems is that they tend to wander and lose focus. Straying from the agenda will lengthen the meeting but not necessarily improve the outcome. If you feel the meeting is drifting, you can simply ask how the discussion ties back in with or interfaces with an item on the agenda. In this manner the committee will automatically refocus and rejoin the agenda. It really works. If you are running a meeting and you find the conversation drifting, pull it back to the agenda. Keep it focused on the problem at hand. If everyone is just talking and essentially saying the same thing, you can end a point by saying: "What I am hearing is that we should..." If you have summarized it correctly, the point is done. NEXT!

Don't Tell Long Anecdotal Stories

One way that committee meetings lose their focus is through the regaling of anecdotal stories or complaints about some other service or problem at the institution. Some stories can last for endless expanses of time and could have been summarized with "the system is broke." If you feel compelled to join the fray of storytelling, be concise. Asking, "how can we fix it?" can often stop the next person from launching into their own rendition of "it is broke." This being said, making a list of the problems expressed can often lead to remedies. So, have a pen out and write down the key problems and be prepared to summarize them if given the chance. Stay focused, keep to the agenda.

Have a Plan

If you are running a meeting, walking into the room with an agenda or a list of items to solve with no plan on what you want the result to be will lead to long meetings and a result you may not be enthralled with. The key here is to do your homework, meet with people ahead of time in one-on-one sessions, establish your plan, and get support. Having others already on board for the plan is sure to help guide it through, shorten the meeting, and allow everyone to leave feeling that they accomplished something.

Consider All Parties

In pushing your plan of action, consider all parties involved. To do this, you will need to ask yourself the following questions: What do I have to gain by the plan? What will others gain/lose if the plan is implemented? What is the benefit to the institution? If you find that all of your plans only benefit you, then you should rethink your behavior.

Have an Ending Time

Committee meetings can go on for a long time. If you are chairing a meeting, have a preset ending time. This will reduce the anecdotal stories, focus the group, and keep the agenda flowing. If you are a member of the committee and not the chair, find out from the chair ahead of time how long she thinks the meeting will run and what time she plans to finish. In this polite way, the chair now knows that you are concerned about the length of time.

Deja' Vu

Your university has existed for a long time and many issues that were problems yesterday and were solved today will probably reemerge in the future. This being the case, if your solution is not accepted, you are likely to have another shot at it in the future. Likewise, if your plan is accepted, it is unlikely to stand as the perfect solution forever. However, if you pay attention, you may be able to prevent your colleagues from reinstating a failed plan that was implemented 10 years ago.

> ▪ Committee work is important and is the driving force of change in most institutions. Do your share, be prepared, be mindful of your colleagues' opinions, choose your battles, and don't worry if it doesn't work out in this decade — you will have additional decades to try again.

NATIONAL SERVICE

As you become established in your career, you will find that journal editors, grants administrators, meeting coordinators, publishers, and the local children's fund may seek your expertise, prestige, and opinions for their programs. Participation in the above services is an honor and a privilege; however, you must temper your enthusiasm for such services with your ability to do your own work.

Reviewing Journal Manuscripts

Manuscript review is the most common form of national service. How much time does it take to review a manuscript? This really depends on how complex the manuscript is and how well the work is performed and presented. The better the manuscript, the less time. On average, manuscripts take between 1 and 2 hours to read, and another hour to write the review. Thus, if you review 2 manuscripts per month, then you will spend almost 2 weeks of your year or 4% of your yearly effort reviewing other people's work. Yikes! Why do it? There are several reasons. The first is that peer review provides a level of quality control to science that in the best of circumstances (most of the time) is free of political influence. This allows science to be the best that it can be. Really. The second benefit is that you get exposed to work that you may not normally read and this expands your scientific range and expertise. The third is that participation in journal review is a form of national recognition of your expertise and prominence in your field.

Your national recognition increases if you are asked to participate on the editorial board of a journal. Participation in such endeavors is considered by your colleagues and institution as an achievement and is usually one consideration (among many) during the promotion/tenure decision. As with any service, serving on an editorial board can eat up a lot of your time. If you are publishing at a steady clip and your lab is humming along, then it would be a good experience and you should accept the position. Be sure to determine how many manuscripts you are expected to process in the course of a year. If you need to publish more papers for your promotion, then ask if you can postpone this honor until next year. In either case, you may want to discuss the appointment with your mentor and chair before you accept it.

For more discussion on reviewing papers, see our *Next Wave* article *To Accept or Reject*, reproduced in *The Extras Section*.

Reviewing Grants

Grant review is one of the most important services that scientists provide. Being asked to review grants is also considered as evidence of your scholarship. As described in *Chapters 3 and 6*, if you are asked to participate in the grant review process, you should strongly consider it. Most of the time junior faculty are only asked to participate on an ad hoc basis. Even so, you will need to carefully consider the amount of time that you will spend. Count on two full days per grant that you receive. Thus, you must plan your time very carefully. One approach is to read the grant in the afternoon or evening and write the review the next morning. This gives you time to let the grant settle in and provide you with some thinking time. Because you may be mentally exhausted after reviewing each grant, you should give yourself some recovery time between them.

> ▪ The advice is to start to review your grants soon after they arrive and try to review them on a regular schedule.

Like editorial board participation, you may want to discuss your participation on grant review panels with your chair or with your mentor. These individuals should be able to let you know if you are on track for your promotion, as they are likely to have sufficient experience with the process.

For a more detailed discussion on reviewing grants, see our *Next Wave* article *To Fund or Not to Fund*, reproduced in *The Extras Section*.

KEEPING RECORDS

As with all aspects of your academic life, keeping continuous records of your activities is important so that when it comes time for promotion or time to find another job you won't forget anything and you won't have to search through your files for some of the information. **Appendix 9-1** has an example of a Service Record Worksheet that you can incorporate into your CV.

> ▪ The bottom line with service is to take it seriously and do your best. Try to be involved in the committees that have the greatest impact on your ability to be a scientist. Lastly, do not become overcommitted to the point of sacrificing your science and your career.

PART III

THE FINISH LINE

As the promotion and tenure clock winds down, the Academic Scientist must organize and prepare for the decision by his/her institution. *Chapter 10* discusses the process, with the specific advice to find out the rules and guidelines at your institution as early as possible. This section will also provide the results of the national survey that was performed for this book. Additional quotations of the respondents are included for the reader.

CHAPTER 10

PROMOTION AND TENURE

"**A**ll I have to do is stay funded and I'll get tenure, right?" Maybe, but your institution may require that you develop a national reputation in your chosen field, too. As you approach your promotion/tenure decision consider the following concept: the decision to award tenure is similar to that of hiring you again. The faculty and the institution are deciding not only whether they want to offer you your job again, but whether or not to keep you around forever. They now have firsthand data on what type of faculty member you are. What you need for promotion and tenure, how you go about preparing your promotion portfolio, and what is considered acceptable for promotion vary from one institution to another. However, there are some key factors on which most institutions base their promotion decisions. These, of course, are scholarship, teaching, and service. This chapter will address promotion and tenure, the elements of a successful promotion portfolio, the presentation of your promotion materials, and the promotion and tenure review process. The following points will be addressed:

- The meaning of promotion and tenure
- Learning the promotion guidelines of your department, school, and institution
- Preparing the promotion packet documentation
- Maintaining your cool during the review period
- Dealing with an unexpected outcome

ASSOCIATE PROFESSOR WITH TENURE

What does it mean to be an Associate Professor? For Academic Scientists, promotion to this academic rank usually coincides with the period of maximum development. Associate Professor *with tenure* is the designation awarded to those individuals who have established themselves in their fields, have participated in the activities of their institutions, have learned how to approach the art of teaching, and are now ready to fully explore and further develop their creative academic genius.

Although promotion to Associate Professor is often accompanied by the award of tenure, a few institutions only award tenure at promotion to full Professor. The T-word is an important aspect of academic life. The dictionary definition of tenure, "permanent status granted to an employee usually after a trial period," provides the most common understanding of tenure. However, the concept can be traced to a set of guidelines that were agreed upon in 1940 by the American Association of University Professors and the Association of Colleges and Universities. The point of tenure is to provide professors with academic freedom in research and teaching. Schools that award tenure have written policies on what tenure means, the limits of tenure, and your rights and responsibilities as a professor. In many schools, the award of tenure equals a job for life. However, seldom does it state that the job carries a salary or the amount of compensation. Thus, while tenure may provide you with academic freedom, it may not pay the rent.

The question most often asked by Assistant Professors about tenure is "what do I have to do to get it?" The answer is simple: perform like a tenured Associate Professor. Promotion and the award of tenure are based on the institution's prediction that you will be a successful faculty member in the years to follow and that you will be a valuable contributor to the goals and plans of the institution. Institutions that award tenure have written their guidelines to define specifically the aspects of teaching, research, and service that are considered in promotion and tenure decisions.

For Academic Scientists in research-centric institutions, scholarship is usually given the highest priority. Your scholarly achievements will be evaluated by your published work, your ability to get funding for your research, your presentations at national meetings, your participation on editorial boards or as a reviewer of manuscripts and grants, and the reputation that you have developed in your field. The latter is almost always judged from the letters of recommendation that your department chair will solicit from knowledgeable individuals in your field. Depending on the priorities of your university, the promotion and tenure decision may be based on whether or not you are the best in your field, in the top 10–20% of your peers, or are at least "a player" in your area of research.

The quality of your teaching may also be given a very high priority in evaluating your tenure eligibility. In some cases, you may have to demonstrate a "national" reputation in teaching to be awarded promotion and tenure for teaching. In other cases, student and faculty evaluations of your teaching skills may be proof of your value to the education of the students at your institution. If a national reputation is required, then publishing a textbook, conducting workshops at national meetings, or participating in the preparation of National Board exams may be considered as proof of your accomplishments.

Service to your institution might also be considered to be a basis for promotion. As noted in previous chapters, service could include running a core laboratory, managing a clinical laboratory, administering an academic program, or providing clinical services to patients. While being promoted based on outstanding service contributions seems reasonable, it is unusual for an Academic Scientist in a research-centric institution to be awarded tenure based *solely* on service. If tenure and promotion were to be based on service, then having a national reputation in service is likely to be important. Evidence of your national reputation in service could be that you are, for example, a leader of a national organization in your field, the editor of a major journal, or are on the organizing committee of a national meeting. As you can see, the opportunities to establish a national reputation in service, separate from a clinical practice, is difficult.

While each of the areas of teaching, scholarship, and service are addressed as separate qualities of your performance, it is the entire package that the peer reviewers evaluate. Some institutions state explicitly in their guidelines that tenure is awarded on the basis of excellence in scholarship only.

Now, let's get started.

THE PROMOTION PROCESS

"My promotion packet will only take me an hour to put together and give to the chair. She'll get the secretaries to make it look sharp before it goes to the review committee." The Assistant Professor who says this is way too optimistic for his/her own good. Would you send in a manuscript to *Top-of-the-Heap Journal* if you had delegated the preparation totally to someone else? Certainly not! So don't entrust the final preparation of your promotion documents to the department staff. Your careful attention to the presentation of your accomplishments is essential because the clarity — or lack of clarity — of your promotion materials can be a factor in determining the outcome of the promotion decision.

The Dossier

One of the keys to putting together an outstanding set of promotion documents is to start the preparations long before the due date. Some aspects of your preparations, such as the ongoing perfection of your Curriculum Vitae (**Appendix 1-1**) and the development of your teaching and service portfolios (**Appendices 7-3 and 9-1**), should start years before you come up for promotion. The advantage of having much of the data already "filled in" for big parts of your promotion packet before you start to finalize the dossier is that you will have time to pay more attention to other important details, such as: "What guidelines am I s'posed to follow?"

Even if you previously reviewed all of the promotion guidelines that you think apply to you (i.e., the guidelines of the university, the school or college, and the department), you should review them again because some aspects of any or all of the guidelines may have recently changed and **no one will remember to tell you about it!** Surprises like this are common, believe it or not. So get hold of all of the appropriate *current* documents.

Guidelines for promotion and tenure usually include not only descriptions of the specific documents that are required in the dossier and the format that they should follow, but may also include a lexicon of words describing the criteria for appointment and promotion to tenured rank. The lexicon could include words like "outstanding," "excellent," "good," and "adequate" and are very important because your chair, the scientists who write your letters of support, and the internal reviewers will use these words in their evaluation of the quality of your teaching, research, and service. The promotion guideline document may also provide general descriptions of the kinds of activities that exemplify each criterion.

For example, the guidelines may state that you must be excellent in scholarship. Excellence in scholarship could then be defined as demonstrating a national reputation for significant contributions in a field as evidenced by:

- Demonstration of independence in research creativity
- Authorship of articles in leading journals
- Continuous success in obtaining extramural research support
- Authorship of review articles, chapters, or books
- Invitations to speak on your research at international or national meetings
- Participation in manuscript or grant review

While there are other areas typically mentioned, these are clearly the most important. The first three items are extremely important and are where problems arise.

Independence

Research independence issues arise when a promotion candidate publishes all of her papers with a senior faculty member at the institution or with her former advisor. How are the evaluators to know who made the key discoveries? If you see yourself in this position, you may want to discuss this with your collaborator and try to "break away" early enough in your career to prevent this circumstance from being an issue. If you can't break away, you should discuss your progress and options with your chair on how you can demonstrate your independence and your contributions to the work.

One way to show independence is by placing your name in the senior authorship positions of your papers. If possible, you should be both first and last author. This is accomplished by having someone in *your* lab appear first and you last. This arrangement is usually interpreted by the scientific community as meaning that the majority of the work was performed in your lab. If you are first and last, your chair can make the argument that you are independent and the driving force behind the work. Additionally, make sure that your grant support is listed on any paper that may raise concerns about your independence.

Publications and Journal Rank

The single issue here is: Have you made a substantial contribution to your field? The answer to this question is determined by the quality and quantity of your publications. While great science can appear in any journal, it is an unfortunate reality that if your work is not in your society-level or better journals, your contributions may not be considered significant or substantial. You will have to rely on an expert in your field to tell your promotion reviewers that your work is great and should have been published in *Top-of-the-*

Heap Journal. Although you may have one big-time paper published, it is possible that your promotion review committee will not consider this enough. Because the time to acceptance of a paper can take between 6 and 8 months (this includes review time, revisions, and re-review), you will need to make sure that your papers are submitted in time for your promotion review. As stated in the survey quote below, publishing your work cannot be over emphasized:

> **Survey Says:** A tenured Full Professor's simple advice to junior faculty: "Publish, publish, publish!"

Continuous Funding

Why should continuous funding be an issue at an academic institution? The simple answer is that research institutions want their faculty to do science and science costs money (money that they do not have). In fact, science costs lots of money; therefore, they want to make sure that you will be able to carry out your research and further the research enterprise of the institution.

The issue of continuous funding has become more difficult over the last 15 years. Up until the late 1980s, federal research grants were often awarded for only three years. This automatically meant that professors would have to demonstrate the ability to renew a grant by the time they came up for tenure, even if they were initially funded after two years of their appointment. It also meant that a faculty member could have more than one opportunity to resubmit a grant prior to the tenure decision. Now, many NIH grants are awarded for five years. Thus, if your first grant starts in your second year of appointment and you come up for promotion consideration in the beginning of your sixth year, you would not have had a chance to renew your funding. One solution to this dilemma is for you to obtain a second extramural grant during this six-year period. A side benefit of this suggestion is that you would have overlapping support and your lab would still be running if you ran into a funding snafu. If you haven't renewed a grant or obtained an additional funding source, then the promotion reviewers will attempt to predict your "fundability" (ability to perform research in the future) from your publication record. One clever approach to managing both your research and career is to time when you get a second grant with your tenure clock. That is, you would like your second grant to begin a year or two before you come up for tenure. This ensures that no matter what the tenure decision is you will be funded. It is also difficult for institutions to turn down someone for tenure when they have several years left on a grant.

As this edition is being written, federal funding budgets have flattened and are not growing at the rate that they did in the early 2000s. This creates another dilemma for the assistant professor in that the ability to get a grant funded has become more difficult. The way around this is to make sure that your grants are submitted to multiple funding agencies. Young investigators have unique opportunities for funding with private foundations that are not available to senior faculty. Do not forget to take advantage of such opportunities. See *Chapter 3* for more details.

Additional Evidence of Scholarship, Teaching, and Service

The other examples of scholarship were presented in the highlighted box at the beginning of this section and included authorship of review articles, invitations to speak at national meetings or at other institutions, and serving as a manuscript and grant reviewer. These are important experiences and demonstrate to the institution that others believe your scholarly achievements to be significant.

Examples of achievement in teaching and service are also likely to be provided in your institution's promotion guidelines. If your institution's promotion process documents do not give explicit information about how the evaluation criteria are implemented, you should talk to your mentors and your chair to get their input. You might also be able to get information about previous successful promotion candidates so that you can see what their achievements were in the areas of scholarship, teaching, and service.

If possible, get permission to see the promotion packet that a successful candidate for promotion prepared. This will help you in putting your credentials together for review. Because the successful candidate whose example you are following may have been promoted under different guidelines, be sure that you follow the *current* promotion packet instructions. Note the way that the packet is arranged, the evidence that is included to support the candidate's case for promotion and tenure, and the thoroughness of the teaching portfolio, as well as the descriptions of service activities. Evidence of achievement in teaching and service often takes more effort to provide than does evidence of research productivity, because funded research grants and publications "speak for themselves." Because faculty must demonstrate some level of achievement in all three areas of academic life in order to achieve tenure, take the opportunity to see how your predecessor in the tenure pursuit presented his or her teaching and service accomplishments.

When your chair gave you the promotion guideline documents, you may have received not only the institutional guidelines, but also those of the department and the school/college. Read everything — twice. And use a highlighter on specific details. If you have a question about anything, ask the chair, and if she can't answer, then ask her who might be able to give you the information. For instance, say that your school's guidelines state that: "evidence of achievement in teaching at the professional and graduate levels should be included in the dossier." What do they mean by "evidence"? Do student evaluations count? If so, then how are these presented in the packet? How about your involvement with graduate thesis committees — does that count? What kind of "achievement" is the promotion committee looking for in this criterion?

In summary, check all of the guidelines that apply to your promotion and tenure decision and get clarification from a knowledgeable source about any parts of the documents that confuse you. Get all of this out of the way long before your promotion packet is due to the chair, because the rest of the process hinges on these "rules of the game."

PUTTING THE PORTFOLIO TOGETHER

Before you do anything else, refer back to your (dog-eared) copy of the institutional guidelines for promotion and tenure to check again on the list of items to be included in your promotion packet (aka, your "portfolio," or your "dossier"). Somewhere in the guidelines you'll find the order in which the documents are to be presented, as well as the prescribed format for documents such as your CV. Go over the list with your chair and/or the department business manager to make sure you have the details right so that you can save time.

> **Survey Says**: A tenured Associate Professor — "It is important to get information [on tenure] ... early. The only reason I know anything about how the tenure process works here is because I eventually served on an executive committee that handled promotions."

It may seem that some of the requirements for your portfolio and the prescribed formatting of the documents are arbitrary. But, if you consider the fact that the university must be careful to handle promotion and tenure processes in the same way for every candidate, then the importance of the esoteric rules becomes clear. The usual components of the promotion packet are:

- Curriculum vitae
- Research plan
- Description of present and past research support
- Reprints of publications
- Teaching portfolio
- Names and addresses of references

Curriculum Vitae

A sample CV is included in **Appendix 1-1**, but the promotion guidelines at your institution will provide the format to be used for your CV for the packet. Some institutions are moving to web-based CVs and promotion packages. While this seems like it's the best thing for the 21st century, it will take a lot of time to fill out the webpages, so plan ahead. Follow the format, spell check everything, and ask your colleagues and your faculty mentor to review it before it is finalized.

Research Plan

As an Academic Scientist, an important part of your tenure and promotion review will be not only what you have accomplished, but what you intend to contribute to your scientific field. You probably don't remember your research description that accompanied the job letter that you wrote when you applied for your current position. Accompanying your application was a research summary and future plan. Your research plan for promotion and tenure should be similar. The key points outlined below should be included in the description.

- Overall summary of research goals
- Summary of accomplishments
- Specific accomplishments, highlighting the novelty and importance of the work
- New ideas/concepts derived from your work
- Future directions — 3-year plan

Because tenure and promotion committees are usually composed of people who are not experts in your field, the research plan should be written so that it is understandable to all scientists, not just those in your field. Try to use as little field-specific jargon as possible. To check on the readability of your document, ask a colleague outside your field read it. If you mention that one aspect of your work was published in *Top-of-the-Heap Journal*, do it only once. That is, brag a little, but not too much. Your external letters of recommendation will help you in this regard. It is important to keep your plan reasonably short. Three pages is about the right length.

Because these plans sometimes go to the outside reviewers, some people worry about revealing their top-secret data and concepts. If you are worried, you may want to prepare and present one plan for the internal review committee and an abbreviated plan for the outside reviewers (if that is within the rules of your institution). If this is not allowed, and you are truly concerned, leave out the top-secret stuff and rest easy.

Description of Research Support

As an Academic Scientist, your record of extramural support is an important factor in the evaluation of your research scholarship. Aside from the obvious indicator of your financial value to the institution, your record of support indicates an evaluation of your research work by peer-review groups in your scientific field. The list of all of your extramural research support, both past and current, might be a required part of your CV, if your institutional promotion guidelines call for it. Alternatively, it could be a separate section of the packet. Either way, you'll be ready with the necessary information if you follow the NIH requirements for their "Other Support" pages, which clearly spell out the components that are pertinent about research support. That is, the titles of the grants, your role on the projects (PI, Co-Investigator, etc.), the dates of the awards, the amount of direct and indirect funding, etc. In addition, a brief narrative description of the projects along with a description of the relationship of each project to other funded or submitted grants or applications is appropriate to include in this section.

Publications

Because your published work is essential to the review of your progress as a scientist, you will have the opportunity to select your most significant refereed publications for your promotion packet. Select from your publications in the top tier of journals in your field and include the publications on which you are first author and those on which are the last. If you are a 'tweener (middle author) on papers that were published in *Top-of-the-Heap Journal*, and if the work is directly related to the research emphasis of your lab, then you should include these papers in your promotion packet, too. Your faculty mentor and your chair will help you choose the best papers to present if you are in doubt. Reprints from the publisher or downloadable PDF files make the best presentation of your publications, but if you don't have publisher's reprints, make the best possible reproductions in the numbers required for your promotion packet. Do not simply rely on the severely abused xerox machine in the back hallway.

Teaching and Service Portfolios

In preparing your promotion packet, be certain to distinguish among the different educational programs that you have participated in (graduate, undergraduate, professional), and be clear about your contributions to each. Don't forget that the postdocs you have trained are part of your teaching record. *Chapters 7 and 8* address the wide range of teaching activities and suggest a variety of ways in which to document your contributions to teaching. Wherever necessary, include in this section cross-references to your publication and research sections of the portfolio to cite the contributions of your postdoctoral and predoctoral trainees.

Whether or not you include inches-thick documents to support your teaching record (that is, documentation of every syllabus of every course you have participated in; or reams of paper documenting individual student evaluations) is a matter for you and your chair to determine. Some chairs go for this kind of thing, but some school-level promotions and tenure committees frown upon it. Do what your chair recommends you do to document your teaching activities.

Appendices 7-3 and 9-1 provide templates for a teaching and service CV, respectively. If you begin keeping these up-to-date throughout your first years as an Assistant Professor, it should be very useful to you by the time you start the preparations of the promotion packet. Use this information to prepare the final version of the teaching and service sections of your packet according to the guidelines and recommendations of your chair, and you should be able to check this part off without too much trouble.

References

The promotion guidelines will state how many letters will be needed from outside references and how many are required from faculty at your own university. The guidelines will also indicate the required rank and tenure status of the reviewers. You can save time for your department chair by giving her a list of good external and internal reviewers who would be able to provide a fair, unbiased, and cogent review of your progress as an Assistant Professor. This will give her a head start on this important part of the review process.

Your list of reviewers can include individuals from current or former collaborators, mentors, or preceptors. Because the letters from such individuals are usually viewed as biased, there must also be a certain number of review letters from other scientists. So, make a list of scientists who have significant standing in your field and meet the criteria of your institutional guidelines for their own rank and tenure status (most guidelines require all of your reviewers to have tenure or its equivalent at their own institutions). You should probably discuss the selection of these reviewers with your chair, your mentor, and your collaborators so that you choose scientists who are: (1) likely to have an appreciation for your research; (2) have successful careers of their own; and (3) are likely to comply with a request from your chair that they write a letter and send it in time to meet the deadlines.

Along with the names and addresses of your proposed reviewers, provide your chair with a very brief biographical sketch of each person, being sure to note whether or not you have any close associations and collaborations with them. For each potential reviewer, note the specific aspects of your research, teaching, or service that the reviewer would be best able to address. Finally, note any significant contributions that they have made to their fields of study and any distinctions (like a Nobel Prize) that they may have.

Most department chairs will request letters from external reviewers that they choose, rather than exclusively using the list of names that you provide. Therefore, you may find it necessary to deal with the fact that there are certain individuals who you believe would give you an unfair and biased review if they were given the opportunity to do so. (In other words, there are scientists out there whom you don't trust, and one of them just might wind up as one of your reviewers.) This is a sticky area, but one that you may need to address. Talk to your chair about this issue so that she can finalize the list of reviewers with this in mind.

Prepare the list of internal reviewers the same way you do the external reviewers, noting in particular the names of faculty who may have firsthand knowledge of your university service activities and teaching successes. Some institutions require letters of review that specifically address the candidate's teaching contributions, so submit some names of faculty colleagues who know how good you are at teaching at each level (graduate, undergraduate, professional, etc.).

Handing it Over

Once you have all of the parts of the packet together that you are responsible for preparing and you are happy with the presentation of the documents, you may want to have the required number of copies made yourself instead of depending on the department's staff to do this. The reason is that from this point on in the tenure and promotion process, "you" and your academic career are represented on these paper documents. You don't want to rely on photocopy mess-ups at this juncture. In any event, check the copies and make sure that all of the pages are included and are legible.

STAY COOL

Waiting with Patience

Okay. Now what? Your promotion packet is turned in and now you have to wait for the outcome as the process winds its way through your institution's tenure review procedures. Usually, the review starts with a meeting of the tenured faculty of your department, who make a recommendation to the chair. Then the chair makes a determination about recommending your promotion to the Dean, who turns over your credentials to a school-level review committee. This committee then makes a recommendation to the Dean about whether or not you should be promoted. If the recommendation is positive, then the review proceeds through the next steps required at the school level, and then on to the university and board levels for approval. Whew!

Obviously, this is not a quick process. With the normal delays in getting outside letters of recommendation and then the problem of convening busy faculty for meetings, the process can take six to eight months. There are several milestones along the way to the final decision that you might hear about, but you are basically in limbo about your future until you receive the letter from the Dean or your chair. You should also recognize the fact that the deliberation process can get hung up at each milestone, either by negative votes of the majority of the reviewers or by their request for additional information.

What should you do in the meantime? First, you should keep working hard on your research. If during the process you learn that a pending grant or manuscript has been funded or accepted for publication, immediately update your CV and ask your chair to get the updated version into the hands of the reviewers. You should also continue to teach, advise trainees, and perform all of your service activities.

What you should not do is complain to your colleagues or students or anyone else about the promotion process — or about anything at all, if you can restrain yourself. Most of your colleagues will understand that you are in a stressful period and will be sympathetic. Despite this fact, you should not look to your colleagues to be your counselors. You should also beware of the tendency of some young faculty to turn to their graduate students and trainees for sympathy during their tenure deliberations. The reason is that, although your students and postdocs may be eager to hear about the complexities of the tenure process, including your offhand comments on faculty politics and its influence on tenure decisions, *they are not part of the process and they are not your counselors, either*. Besides, you don't want to turn them off to the idea of a career in academic science, do you? Instead, let them think of you as the next tenured Associate Professor.

Waiting in an Aggressive Manner

Some faculty believe that it is necessary to demonstrate their worth by seeking and finding employment at another institution. This tactic is not uncommon and has its benefits and disadvantages. The beneficial edge, of course, is that if you can procure an offer from an equal or better institution (as determined by however one determines this sort of thing), you will be able to say to your chair, "Hey, I've got a job offer at Big Time U at the Associate Professor level, and they say that they will give me tenure." Your chair may relay this information directly to the committee evaluating your package. Institutions in general do not want to lose faculty and therefore this strategy has been used many times. The other benefit is that if the decision looks like it is not going to go well for you, you will have another job waiting in the wings.

The disadvantage to this course of action is that it will consume a lot of your time and energy and ultimately reduce your productivity. In a worst-case scenario, your department may now find it easier to say, "maybe things will work out better for Dr. Starr at Big Time U than it did here"; or "thank goodness, she got another job." So, goes this academic game of chicken.

WHAT TO DO IF IT DOESN'T GO YOUR WAY

At each level in the decision-making chain there are administrative, academic, and fiscal considerations that must be taken into account regarding all promotion and tenure actions. Tenure, after all, implies a long-term financial commitment to keep you employed. This is why tenure is *valuable!* Because there are many factors that must be considered, it is possible that, despite your success as a scientist, scholar, and educator, some factor beyond your control may have a negative effect on your tenure decision at your current institution.

If you are turned down for promotion and tenure, then immediately talk to your chair about your next steps. What action you take is dependent upon a number of factors, such as whether or not you were turned down at the department, school, or university level. It is sometimes possible to appeal due to the nature of the decision and how the process was followed. It may also be possible to switch to a non-tenure-track position if you still desire to remain at the same institution.

Try to be cooperative in every way and try not to let disappointment undermine your confidence in yourself and your relationships in the department. Remember that if you leave this job you may want to maintain the good relationships that you have built in the department and the school.

What to Do Next

Depending on the timing of the decision and when your position expires, your options change. You now have the opportunity to ask yourself the same question that was posed at the beginning of this book. What type of scientist do you want to be: academic, industrial, or do you want to do something different? This is an important question. The answer to this question may now be different.

Seeking your Next Academic Position

If you still want to pursue academics, the first thing to do is take an honest look at your CV. How many papers did you publish in the last 5 years? What journals were they in? How many extramural grants did you procure and are they still active? If there are some weaknesses that you perceive, now is the time to take action and get your CV into fighting form.

If publications were a problem, then it is time to go over your experimental data and look to complete as many minimal publishable units as possible. Write up the papers and submit them to your society-level journals. This will serve the purpose that your CV will be boosted as soon as the papers are submitted and it will be more attractive to other institutions who would might say, "I can't believe that Research Utopia University let Dr. Starr leave. Look at these new papers!"

If funding issues were your problem and you have lots of publications in society-level journals, then you need to determine if it was your timing on the grants (i.e., your score was close and you need another round) or whether your grantsmanship skills are not quite honed yet. If the former is the case, then all you need to do is to address the critique of the grant and resubmit it. You will be able to transfer the grant that you get to your new institution. If the problem lies in your granstmanship skills, then you need to seek help. As mentioned in *Chapter 3*, program directors at the NIH are willing to help. Call them and talk to them about your grant and how to improve your approach to presenting your science. Maybe you need to be in a different study section.

If you haven't done so already, try to target your grant applications to the private funding agencies. This may be a way in which you can get some money to make hiring you a plus for your new department (remember *Chapter 1*).

Sending Out Letters and Applications

As when you started your quest for a job in academia, start looking through the job ads and begin to make your list of places where you would like to apply. At this point in your life, your knowledge of the science community has matured and your views are likely to have changed. You may no longer want a position at Research Utopia University or Big Time U, because you now realize that you really desire a position that will have a better mix of teaching and a reduced emphasis on research. As such, a position at Teachem Right State may be for you. Use your knowledge base in choosing where you apply. Your CV is really ready, and with the additions that you are making to your publications and funding, you should be ready to roll.

Seeking Alternative Careers

If you decide that you would rather do something else in your career, you will need to formulate a plan of attack. If you don't know what you truly want to do, then you need to find out what is out there. It is common for institutions to sponsor scientific career alternative workshops and seminars. Some of the big scientific society meetings also sponsor such sessions and may even have job placement services that could be of interest to you. Yes, now is the time to be aggressive and seek out all the information you need on your newly chosen path. Because science training teaches ways to logically approach and solve problems, there are lots of opportunities. Call your old buddies who went to these positions and ask them what it is like, how well they like their jobs, and if there are any openings.

> ▪ Regardless of which direction you choose, take charge of your life and look forward to your change whether it is your promotion to tenured Associate Professor where you are now, or somewhere else. With a positive attitude and a smile, your new job will be great and you will be happy. You will soon hear them say: "Hey, there's the new kid on the block!"

CHAPTER 11

SURVEY SAYS ...

"**D**on't be afraid to ask advice of colleagues, both at your own institution and, perhaps more importantly, at others, since the latter group is much larger and just as accessible in this day of near instantaneous conversation." This quote is from a tenured Associate Professor of cell biology, who was among the 102 Academic Scientists who responded to a web-based survey that was conducted for this book in the Fall of 2001. This recommendation, on the importance of seeking advice from other faculty, is one of many comments that were received from these anonymous yet helpful colleagues from a number of research universities. This chapter provides a summary of the results of this survey and highlights of the advice that these faculty members have to offer from their own experiences as Academic Scientists. After a review of the comments some four years later, little has changed. The advice is still valid.

THE SURVEY

Faculty from medical school basic science departments at 17 of the top-ranked US research institutions were surveyed. The survey questionnaire, which appears at the end of this chapter, sought the views of basic science faculty on issues about establishing and succeeding in research careers and progressing to tenure in an academic institution. The questionnaire was organized into the following four sections:

1. *Beginning Your First Faculty Position* included questions about the importance of factors like the amount of start-up funding and the ability to hire good lab staff. This section also asked about satisfaction with the respondent's schools' doctoral programs.

2. *Tenure Policies and Procedures*. This section asked about the tenure policies at the respondent's institution and the evaluation of faculty productivity.

3. *Tenure and Promotion Factors* included questions about the importance of teaching, research, and service activities in gaining tenure at their institutions.

4. *Background Information*. This section asked the respondents their rank, tenure status, year of first appointment, etc. These responses were used to sort the data.

Space was also provided in the survey for the respondents to write their comments and give advice about achieving career success to up-and-coming Assistant Professors. Happily, a number of the respondents took the time to give their advice. As you no doubt noticed, these comments are included in the shaded boxes throughout this book. Select comments are also compiled below.

Responses were received from Assistant Professors, seasoned Associate Professors, and worldly-wise full Professors, providing a variety of viewpoints across the range of the academic ranks. The breakdown of the ranks our respondents turned out as follows: Assistant Professors, 31%; Associate Professors, 23%; and Professors, 46%. Of these, 72% were tenured, 26% were tenure track, and 2% were not on a tenure track at the time of the survey.

The results of the survey were analyzed using the Statistical Package for Social Sciences (SPSS). Analysis of variance was used to assess the individual items in regard to differences among ranks and tenure status. Scale reliabilities were in the range of alpha = 0.6275 to 0.8482. The three non-tenure-track respondents were not represented in the analyses because of their small number.

The information about the rank and tenure status of the respondents was an important parameter in the evaluation of the survey. The perspectives of faculty members about what it means to be an academician change with the additional responsibilities that senior faculty are given in their institutions. Because tenured professors and associate professors are the people who do much of the work in the process of promotion/tenure evaluation of Assistant Professors, they gain a finely tuned perspective on factors that relate to job performance. So, the survey comments from these senior faculty could be considered the inside "scoop" for junior faculty from those with first-hand information about promotion/tenure review.

Following are highlights of the survey responses regarding the "academic deltoid" of research, teaching, and service.

Research Performance

All of the survey respondents work in research-intensive, medical school basic science departments, so it was not surprising that there was strong agreement among them about the overriding importance of research productivity to career success. The respondents were in agreement that the essentials for progression from Assistant to Associate Professor are extramural grant funding and a number of first- and last-author publications in prestigious journals. Because the actual number of publications, the levels of prestige of the journals, and impact of the publications in their field are norms that are set at individual universities and vary by scientific field of study, these specifics were not addressed by the survey.

Even though it is a given that success in research is a job requirement for academic scientists in basic science departments, the survey group had definite opinions about the importance of the factors that can enhance success. Following are some highlights of their opinions on start-up, grant funding, and papers:

Start-Up

- 94% of respondents believe that hiring and training **productive laboratory personnel** is necessary to research success for a junior faculty member.
- The respondents put less emphasis on the **square footage of the lab**, with over 30% saying that size didn't matter. Only 7% thought that the square footage of the lab was "very important."
- 94% of Assistant Professors thought that the **amount of start-up funding** was either important or very important, with 62% of full Professors in agreement on this point.
- Access to **shared research facilities** was important or very important to 65% of respondents.

Funding

- Fully 100% of respondents found that having **NIH RO1 grant support** at the time of promotion was "very important" (93%) or "important."
- Interestingly, **non-NIH R01 funding** was not as highly valued, with 23% citing such extramural support from other federal or foundation sources as inconsequential or not at all important.
- 90% of the respondents said that having **renewed an existing grant** at the time of the tenure decision is important.

Papers

- 97% said that the **number of first- and last-author papers** is very important/important for promotion to tenured rank.
- Only a quarter of respondents thought that **middle authorship** on a publication is important. 42% of the respondents say that being a 'tweener is not at all important or inconsequential.
- The **prestige of the journal** where the publications appear was emphasized by 92% of the faculty who completed the survey.

In summary, the survey respondents from the top-ranked research medical schools recommended that junior faculty emphasize hiring the right people to staff their research labs. Junior faculty should procure NIH RO1 funding as soon as possible, so that they can get this first grant renewed by the time they are considered for promotion. The junior faculty should emphasize publishing results as first- or last-author papers in prestigious journals.

Teaching Performance

University faculty in medical basic science departments may not teach as much as their colleagues in other parts of the university do, but they probably won't be promoted unless they teach well. The responses to the teaching section of the survey show the following trends:

- In regard to **classroom teaching**, the quality of the junior faculty member's teaching clearly takes precedence over the quantity. Full Professors were more likely to emphasize the quality factor in the promotion and tenure evaluation.

- 42% of the survey respondents said that **teaching awards** can help in the tenure and promotion evaluation, but the rest don't believe it is important at all to list such an award on the CV of a junior faculty member. Because the respondents have already noted that the quality of teaching is important, this result is likely a criticism of institutional teaching awards, and the factors that lead to the selection of the recipients. In other words, such awards may not necessarily signify teaching quality.

- 26% of the survey respondents say that participating in **writing textbooks** could help enhance a junior faculty member's teaching CV.

- As far as promotion is concerned, **serving on the dissertation committees** of other students was not considered much of a factor: more than half said that being a member of such a committee is either not at all important or completely inconsequential to the promotion evaluation of a junior faculty member.

- The survey respondents would like to have **better access to doctoral students**: less than one-tenth of them cited being "very satisfied" with their access, and a quarter of the respondents were not at all satisfied.

- Interestingly, the faculty were satisfied with respect to the quality of the **graduate students who joined their labs** — once they found them. In fact, nearly three-quarters noted that they were satisfied to very satisfied with their doctoral trainees.

Service Activities

Research universities have many opportunities for faculty to contribute to the governance of the institution. But, junior faculty should probably consider waiting to help out in a service assignment until after they are promoted. As one tenured Associate Professor puts it: "Delay service on committees until grant funding and publications are achieved." Some service, though, is a requirement for promotion to tenured rank in universities, but it's important to choose the service work to maximize the value of this time commitment.

- Less than 50% of the survey respondents cite **service to the department and university** as important. When asked about specific types of service, these academic scientists found that there are certain jobs that warrant the devotion of their time. For instance,

- 42% say that **participating in the governance of graduate programs** at the institution can be important to the promotion and tenure evaluation, and

- 48% say that directing or coordinating the work of a **core lab facility** could enhance the junior faculty member's CV.

- National service is more valuable to our survey respondents, especially service that relates to research and scholarship. Over 80% believe that membership on an NIH **study section or peer-review committee** of another funding entity was important for tenure and promotion.

- 80% also say that membership on the **editorial board of a journal** is important for career success.

In summary, the advice from our survey respondents is to concentrate on research, scholarship, and publications before getting involved in service activities. Then, maximize the service impact while enhancing the research standing by serving on peer-review groups and editorial boards. This could be considered a "two-for-one" deal — counts as service and works toward research goals.

QUICK TIPS ON CAREER SUCCESS

The survey participants were generous in offering their narrative advice through the survey form. Following are some of their "quick tips," which are some of the brief comments some of the respondents wrote:

> "Focus on important research problems. Be original. Get your research articles out in a timely manner. One a year for 4 or 5 years is better than 4 or 5 the year before promotion. Attend national meetings with poster contributions."

> "If you're at a research-intensive university, focus exclusively on research. They'll pay lip service to the importance of teaching, serving the graduate program, etc., but when it comes down to it, the only important thing is the amount of grant dollars your research brings in."

> "Do clever science."

> "Publish. Publish. Publish."

> "[Develop] support from the chairman in getting an R01 grant and good quality graduate students and postdocs."

> "[The keys are] scientific luck, clear and interesting scientific goals, hard work, respect and nurture for junior collaborators."

> "[Get a] well-funded, nationally/internationally recognized research program; [do] good teaching; forget all committees!!"

> "Get grants and forget teaching committees and other department duties if at all possible."

> "Teaching, service at both the national and institutional level, having an ongoing research program leading to research publications in peer reviewed journals, and having graduate students that are productive are all necessary for promotion."

"Scholarship is paramount. This means identifying and attacking questions of high general interest, not just any questions that CAN be addressed."

"Work hard, be effective, seek and take advice from successful academics, be creative, be cooperative and willing, develop a network with successful scientists/colleagues."

"Focus, focus, focus [on] NIH grants."

"A cheerful disposition, decent job on assigned committees and teaching and excellent research are about all that mattered."

"Grant funding is the major factor. This requires not just good ideas, but a record of publications. Delay service on committees until this is achieved."

"Important factors:
    Clear communication with your division head and chair
    Focus on career building
    Keep research focused
    Focus on NIH funding
    Quality publications
    Join relevant graduate faculty
    Teach."

As you can see from this list, each comment represents a faculty member's view. The consensus of these comments is to do the best science that you can, be a good teacher, and contribute to your institution.

THE ACADEMIC SCIENTIST SURVEY ON TENURE

(CONDUCTED SUMMER 2001)

This survey is part of a project on the views of basic science faculty regarding the establishment of their research careers and the progression to tenured rank. None of the data provided by this survey will be identified by name or institution.

Please take a few moments to answer the questions below. We welcome your comments related to specific sections of this questionnaire in the space provided and at the end of the survey. Thank you.

I. Beginning Your First Faculty Position

How did the following factors affect the establishment of your research lab during your first 2 years as an Assistant Professor?

- **Amount of start-up funding**.
  - ❏ Not at all Important
  - ❏ Somewhat Important
  - ❏ Neutral
  - ❏ Important
  - ❏ Very Important

- **Size (square footage) of the laboratory.**
  - ❏ Not at all Important
  - ❏ Somewhat Important
  - ❏ Neutral
  - ❏ Important
  - ❏ Very Important

- **Access to shared, "core" facilities (e.g., transgenic mouse: DNA microarray, etc.).**
  - ❏ Not at all Important
  - ❏ Somewhat Important
  - ❏ Neutral
  - ❏ Important
  - ❏ Very Important

- **Hiring and training product lab personnel.**
  - ❏ Not at all Important
  - ❏ Somewhat Important
  - ❏ Neutral
  - ❏ Important
  - ❏ Very Important

- **Advice from established faculty investigators.**
  - ❏ Not at all Important
  - ❏ Somewhat Important
  - ❏ Neutral
  - ❏ Important
  - ❏ Very Important

Comments:

I.a. Graduate Students

- Access to good doctoral students is often noted as an important factor in the successful development of academic research labs. Please rate your satisfaction with your access to doctoral students.
  - ❑ Not at all satisfied
  - ❑ Somewhat satisfied
  - ❑ Neutral
  - ❑ Satisfied
  - ❑ Very Satisfied

- Please rate your satisfaction with the outcome of having doctoral students involved in your research programs.
  - ❑ Not at all satisfied
  - ❑ Somewhat satisfied
  - ❑ Neutral
  - ❑ Satisfied
  - ❑ Very Satisfied

II. Tenure Policies and Procedures

Most institutions have published polices and procedures regarding tenure and promotion for departments, schools and the university as a whole. Please note your knowledge of these policies and procedures for each of these levels of the institution.

	KNOW THE POLICIES AND PROCEDURES Yes/No	KNOW HOW TO GET THIS INFORMATION Yes/No
Department	Yes/No	Yes/No
Medical School	Yes/No	Yes/No
University	Yes/No	Yes/No

- When did you being to evaluate your own productivity in relation to your candidacy for promotion to tenured rank?
  - ❑ Before starting your faculty position
  - ❑ After the first year
  - ❑ After three years
  - ❑ Just prior to starting the tenure process

III. Tenure and Promotion Factors

Please rate the importance of the following factors on the likely success of gaining tenure and promotion in your department and university.

Teaching

- **Quality of tenure/promotion candidate's teaching.**
  - ❏ Not at all Important
  - ❏ Somewhat Important
  - ❏ Neutral
  - ❏ Important
  - ❏ Very Important
  - ❏ Don't Know

- **Teaching more than other faculty members do.**
  - ❏ Not at all Important
  - ❏ Somewhat Important
  - ❏ Neutral
  - ❏ Important
  - ❏ Very Important
  - ❏ Don't Know

- **Membership on doctoral dissertation committees.**
  - ❏ Not at all Important
  - ❏ Somewhat Important
  - ❏ Neutral
  - ❏ Important
  - ❏ Very Important
  - ❏ Don't Know

- **Winning teaching awards.**
  - ❏ Not at all Important
  - ❏ Somewhat Important
  - ❏ Neutral
  - ❏ Important
  - ❏ Very Important
  - ❏ Don't Know

- **Participating in writing textbooks.**
  - ❏ Not at all Important
  - ❏ Somewhat Important
  - ❏ Neutral
  - ❏ Important
  - ❏ Very Important
  - ❏ Don't Know

Research

- **Quality of candidate's research program.**
  - ❏ Not at all Important
  - ❏ Somewhat Important
  - ❏ Neutral
  - ❏ Important
  - ❏ Very Important
  - ❏ Don't Know

- **Having NIH, RO1 grant support at the time of tenure/promotion decision.**
  - ❏ Not at all Important
  - ❏ Somewhat Important
  - ❏ Neutral
  - ❏ Important
  - ❏ Very Important
  - ❏ Don't Know

- **Having other types of federal or foundation grant support.**
  - ❏ Not at all Important
  - ❏ Somewhat Important
  - ❏ Neutral
  - ❏ Important
  - ❏ Very Important
  - ❏ Don't Know

- **Having renewed grant support at the time of the decision.**
  - ❏ Not at all Important
  - ❏ Somewhat Important
  - ❏ Neutral
  - ❏ Important
  - ❏ Very Important
  - ❏ Don't Know

- **The number of "first-author" and "last-author" papers on candidate's CV.**
  - ❏ Not at all Important
  - ❏ Somewhat Important
  - ❏ Neutral
  - ❏ Important
  - ❏ Very Important
  - ❏ Don't Know

- **Value of "middle authorship" on publications.**
  - ❏ Not at all Important
  - ❏ Somewhat Important
  - ❏ Neutral
  - ❏ Important
  - ❏ Very Important
  - ❏ Don't Know

- **Prestige of the journal where publication appear.**
  - ❏ Not at all Important
  - ❏ Somewhat Important
  - ❏ Neutral
  - ❏ Important
  - ❏ Very Important
  - ❏ Don't Know

Service

- **Membership on the editorial board of a journal.**
  - ❏ Not at all Important
  - ❏ Somewhat Important
  - ❏ Neutral
  - ❏ Important
  - ❏ Very Important
  - ❏ Don't Know

- **Membership on a Study Section at NIH or another extramural funding agency or foundation.**
  - ❏ Not at all Important
  - ❏ Somewhat Important
  - ❏ Neutral
  - ❏ Important
  - ❏ Very Important
  - ❏ Don't Know

- **Clinical service (patient related) activities.**
  - ❏ Not at all Important
  - ❏ Somewhat Important
  - ❏ Neutral
  - ❏ Important
  - ❏ Very Important
  - ❏ Don't Know

- **Directing or coordinating a core lab facility (e.g., transgenic mouse, DNA sequencing, etc.).**
  - ❏ Not at all Important
  - ❏ Somewhat Important
  - ❏ Neutral
  - ❏ Important
  - ❏ Very Important
  - ❏ Don't Know

- **Candidate's participation in the governance of a graduate program at the institution (e.g., admissions committee, curriculum committee, etc.).**
  - ❏ Not at all Important
  - ❏ Somewhat Important
  - ❏ Neutral
  - ❏ Important
  - ❏ Very Important
  - ❏ Don't Know

- **Service to the Department and University (e.g., serving on Faculty Senate, coordinating departmental seminar program).**
  - ❏ Not at all Important
  - ❏ Somewhat Important
  - ❏ Neutral
  - ❏ Important
  - ❏ Very Important
  - ❏ Don't Know

IV. Background Information for Statistical Analysis

- Year obtained PhD or MD:
- Current Rank:
- Tenure Status:
- Field of Study:
- Number of years of postdoctoral study prior to first Assistant Professor position:
- Year of first Assistant Professor appointment:

- **Approximate amount of start-up funding of your first Assistant Professorship.**
  - ❑ <$150,000
  - ❑ $150,000–300,000
  - ❑ $300,000–500,000
  - ❑ >$500,000

Thank you for your participation in this survey. The results will be kept confidential and used to provide general information for beginning Assistant Professors in the life sciences.

Please take this opportunity to help Assistant Professors establish their research careers in the life sciences by providing some information about what you have found to be important for successful academic scientists.

PART IV

THE EXTRAS SECTION

(DAILY DOUBLE)

Since the publication of the first edition of *Academic Scientists at Work*, the authors have contributed to *Science*'s on-line *Next Wave* Career Development Website (http://nextwave. sciencemag.org/cdc/index.shtml) as columnists under the banner title of this book. To date, 12 articles were published. Like this book, the 12 essays focus on problem areas that junior (and senior) faculty encounter. While some of the essays cover topics and provide advice that is given in the first 10 chapters of this book, the *Next Wave* articles present a more detailed and thorough discussion of the subject matter.

To provide you with this additional information without having to log on, we reproduced our 12 articles in the pages that follow.

The following article on preparing for your job talk appeared on-line in *Science*'s *Next Wave* on December 10, 2004. The PowerPoint presentation referred to as a PDF below is included in the CD as Appendix 1-3.

Academic Scientists at Work:

THE JOB TALK

10 DECEMBER 2004

"That new Assistant Professor candidate with the curly red hair had some interesting data, but I am not sure he really got his main point across," you say to your tenured colleague just after the candidate's job seminar. Your colleague, Dr. Ben Harawhyle, responds: "Yeah, I think he may have a nice system, too, but he ran out of time and ended just when he was getting to the cool results. Also, his slides were hard to read, especially from where I was sitting." "Well," you respond, "we have some more candidates — I'm sure one will be better." That casual exchange adds up to a job-talk disaster for the candidate with the curly red hair. Don't let it be you!

If you want to win the race, you need to present what the search committee, department chair, and all the department faculty need to see and hear to motivate them to offer you a position. Chances are the position will be in a department with faculty members who have varied research interests, all of whom have some stake in the hire. Hence, your audience will be a complex mix of scientists with distinct and diverse standards. While this sounds challenging, good organization and a clear idea of what is expected will help you in your quest for the dream position.

This article will discuss what you need to present in your job talk, how to organize it, and how to prepare your slides. Dos and Don'ts of job talks are included for good measure. This link contains a PDF of a PowerPoint slide presentation that expands the discussion and provides some examples of color combinations that work, and some that don't!

The Job-Talk Audience

One way to get a handle on the problem of gauging your job talk for the whole audience is to give consideration to the different interests of the group. Get to know your audience ahead of time. Learn something about the members of the department, especially the chair and members of the search committee. Get to know their work. This will help you, not just in preparing your job talk, but in the one-on-one interviews once you arrive on campus.

Once you've got a fix on the likely audience, divide it up into several groups: those who are interested in your field; the search committee and the department chair; and, finally, those who are not in your field and who therefore need to be convinced of the importance of your work.

It's obvious that you have to impress the search committee, and it's always a good idea to impress the chair of the department. What may not be so obvious is that you also have to impress the faculty who are not especially interested in your research. This group has an interest in the hire because it will affect the department's chemistry, and because they want a candidate that's a good "fit" in a broad sense. They also can have an important influence on the decision; if several candidates are relatively equal, the influence of this group may give you the edge. Convincing this latter group of the value of your work while providing the details the experts require is the challenge.

What Do You Need To Present?

You need to tell an intriguing story about your research. Everyone in the audience has to understand it, each at a level appropriate to them. This means that the background and introduction to the problem should be simple and your description of the questions being investigated should be more general than specific. The methodology should be explained in such a way that everyone can follow along and see the logic and the beauty of your approach. You need to present beautiful, crisp data that are well controlled and support your conclusions. Finally, you need to tell your audience where the work is likely to take you — or you to take it — over the next 2–3 years.

Organizing the Talk

Job talks have a specific order. Everyone begins with a thank you to the people who invited them and perhaps a joke or short humorous anecdote that doesn't offend anyone or anything — living or dead. If you are not funny — really funny — skip this part. When this is done, you need to get on to the introduction and objectives. Some people — us included — like the idea of having a brief general introduction to the field, followed by a narrower (but still general) account of the area of study you are working in. This allows you to then state the objectives of the talk as they relate to the field's broader issues.

Objectives can be presented many different ways; we'll mention the two ways that we think have the greatest impact. The first way is to provide your main conclusions as objectives: "Today I will demonstrate that: (1) long distance chromatin loops form between distal and proximal gene regulatory regions; (2) loop formation is dependent on induced sequence specific transcription factors; and (3) loop formation is required for gene expression," and so on. The second way is to present a set of questions that you intend to address ... which amounts to the same thing, since it's easy to change the above statements into questions. For example, "(1) are long distance chromatin loops formed between distal and proximal gene regulatory regions?" and so forth. Take the approach that makes you most comfortable.

After the objectives and the introduction comes the meat of the talk, the presentation of the data and conclusions. Our advice is to make sure that your audience continues to see the connection between the results you're describing and your research objectives. Never let them forget why you did what you did, and why you're telling them about it. Since most of us have had lots of practice presenting data and results to a variety of audiences, further discussion here is not necessary.

Before we get on to the end of your job talk, let's present some Dos and Don'ts.

The ASAW Top-Ten DO List for a Job Talk:

1 Dress professionally — appearance is important everywhere, especially when you are the center of attention.

2 Practice and time your talk so that it concludes without interruption after 50 minutes — this means you must limit the number of slides. Decide ahead of time what parts can be cut if you find yourself interrupted frequently and the talk runs long. But you don't want to

design your talk to be too short; if you don't get many questions, it may seem like you don't have enough results to fill up an hour's presentation.

3 Try to relax during your talk and show everyone that you are really enthusiastic about the science you are presenting. You care about your work; don't keep that a secret. If you get nervous presenting to strangers, it may be helpful to: (a) increase your confidence by practicing your talk a lot; (b) memorize your opening sentences so that you are confident in your start; and (c) have planned and memorized segues to link the sections of your talk.

4 Slides must emphasize content, not form. If the slide is distracting, no one will follow the data.

5 Fancy-dancy animation or movies and other visual effects must work the first time on your laptop or whatever hardware you end up using; are you sure that the presentation you prepared on a PC will work properly on a Macintosh, even if it's running OS9.236? So, find out what the video setup is before you leave home. If you can't be sure, don't use the fancy-dancy technique at all.

6 Have a backup file (CD) for your talk in case of a computer malfunction.

7 Data slides should be neat, orderly, and error free.

8 Slides that contain the data of others should be acknowledged clearly on the bottom of the slide, and noted in your talk. You must not appear to be taking credit for the work of others, even inadvertently.

9 You must reserve time in your talk for your best results and coolest data.

10 Appropriately and briefly acknowledge your colleagues and collaborators.

The ASAW Top-Ten DON'T List for a Job Talk:

1 Do not speak too quickly, softly, or unclearly. You really want your audience to follow your talk.

2 Do not use red lettering on blue or black backgrounds, even if your boss does. See the accompanying examples of PowerPoint slides. Check out all of your slides in a conference room with most of the lights on so you can see how clearly they project.

3 Do not lose control of the time because of questions from your audience. If you will get to a question later, say so, politely. If not, try to answer the question clearly and briefly. If the questioning continues, politely let the audience members know that you will be happy to discuss these terrific ideas at the end because you have limited time and a lot more to present. If the inquisition doesn't end there, you might try something like, "Many of these issues have already been published by our group and Dr. Doitright's group." Do your best to focus on answering the questions without providing Dr. Inquisition with loose ends or speculation to ask more. Sometimes a response that involves a

"we haven't, but need to do an experiment to test your point" will quiet Dr. I.

4 Do not forget that you are on stage and need to keep everyone focused on the science. Do not fidget, gulp down water, pause at length, gaze at the floor, or keep your back to the audience the entire time.

5 Do not wave the laser pointer all over the place; keep it steady on the result of interest. If you are nervous and your hand shakes, rest your laser hand on the podium edge to steady it.

6 Do not ignore your controls or other data on a slide. Remember, proper controls convince everyone that your experimental interpretation and results are correct. If you have data on a slide you are not going to talk about, remove it. It is a distraction.

7 Do not get carried away with wild speculation or over-interpretation unless you have data to back it up (in which case it's not wild speculation or over-interpretation...).

8 Do not patronize the audience. Don't answer questions with "that's a really interesting question," unless you hadn't thought of it before and you truly think it's a great question. You don't want to come off as insincere.

9 Do not use unnecessary jargon in your talk. If you must use some jargon, define it and keep it to a minimum.

10 Do not say anything negative about your lab, labmates, mentor, institution, collaborators, or other scientists (even if they refused to provide you with reagents). This admonition extends, by the way, to one-on-one interviews.

Ending the Talk

As the talk nears its end, you will need to spend some time on what you plan to do once you accept the job offer. Your new department will want to know what lines of research you can bring with you and what you will have to leave to your mentor. If all the cool stuff is staying and you are moving on to a new field, this will not look good, especially if your "job talk" was on your previous work and not on what you intend to do next. You will need to discuss this with your mentor before the job talk, so that you can address this issue if it comes up. It probably will, either in the job talk or in one-on-one interviews.

At the end of the talk, it's a good idea to present your future goals in the form of specific aims. This provides the audience with a glimpse of what your first grant application will look like. The aims should be distinct, yet related, but not necessarily dependent on each other. There should be no more than four, with three being ideal. The aims, like your objectives, can be in the form of statements or questions. Alternatively, if you set up your talk in a manner that allows you to state your future objectives as hypotheses, this approach can be very effective as well.

One area of concern for your audience is what you did and what everyone else in the research group did. This is especially true if you come from a large lab group. To make it clear, use the first person ("I" and "me") where it's appropriate, and acknowledge your colleague's/collaborator's work throughout the talk and/or have their names on the slides where their data were presented. The list of acknowledgments at the end needs to include only those that contributed to the work. If your work was supported by a grant that you procured, you should acknowledge their support. This reminds the faculty that you are capable of getting funded.

Finally, take advantage of opportunities to listen to job candidate talks at your institution. Instead of just listening to the science being presented, pay particular attention to the structure of the talk, how the candidate presents her future goals, how he managed his time and answered questions. Mistakes you notice are mistakes you can avoid in your own talks.

In summary, the job talk is the most important part of the interview process. You were chosen for the interview because of your science and experience, so there is little reason to be very nervous. Practice your talk in front of your advisor, other faculty, and/or fellow fellows so that you can get direct feedback.

Good luck. Knock 'em dead.

The following article appeared on-line in *Science*'s *Next Wave* on February 11, 2005.

Academic Scientists at Work Negotiating a Faculty Position:

GREEDY GUS: NEGOTIATING AN ASSISTANT PROFESSORSHIP

11 FEBRUARY 2005

"**C**an you believe it? Dr. Gus, the immunology faculty candidate with the curly red hair has asked for season's tickets for the Braves in his start-up package," the chair of the new assistant professor search committee announces at the faculty meeting. "I heard that he is using our offer to up the ante at South-by-Southeast U, responds Dr. B. Leaveitornot," to which the department chair quickly responds, "I don't know or care about South-by-Southeast. We are a better fit for his research program, our offer is very generous, and after three rounds of discussion and visits he can take what we offer or not. I will give him until next Thursday." Assuming that Gus really wanted to work at Firstchoice University, instead of South-by-Southeast, he may have blown the deal.

Negotiating a job is similar to playing a hand of poker: the stronger your hand — your credentials — the more you can demand. The trick is to know what aspects of the position are negotiable and what the limits are; otherwise, you may find that the offer has folded. This article will discuss the issues at stake in academic science research positions and offer some suggestions on how you could approach your own negotiations so that you get the job you want and the start-up package you need.

The first and most important key to any negotiation is to know your bottom line. Your prospective employer — the Department Chair — knows *her* bottom line, because she has limited resources and space. That puts you at a disadvantage, unless you know yours as well. To figure out what *your* bottom line is you must first answer these critical questions:

- How good a fit is the department for my research program?
- What size lab do I need?
- How much "start-up" money do I need to establish my research program?
- How much time am I willing to devote to teaching?
- How much time can I afford to devote for service-oriented duties?

The first question asks you to decide whether this would be a good department for you ... and therefore how many other things you are willing to sacrifice in order to end up there. If it's not a good fit, then you have two choices: (1) try to negotiate some aspect of the offer that will make the position more attractive, or (2) look elsewhere.

If this department is your first choice, then you don't want to blow the deal so you will need to be careful to ask only for what you really need to be successful. You may consider asking for things that are not essential for success but would make life better. However, you may have to pass on some of these during your negotiations. At least you know what items/issues you will fold your hand on. As you negotiate, tally your points. As a rule of thumb, you can expect to win roughly four *important* points of negotiation in your final offer. There is no exact number, but no one likes a Greedy Gus, so choose wisely. Only negotiate hard for the things you know you really need.

If the department is not ideal but has some aspects that would make it a good place to work, you can probably afford to push harder for items and issues that would make the position better. Such items may include special equipment that you normally would share, slightly larger space, larger start-up funds, reduced teaching, bigger computer monitor, and others. If you really don't want the job, politely say so and move on. It really is not worth the headache and your time to negotiate for a position you do not want.

Start-Up

Let's assume the department is a good fit. What's involved in negotiating the start-up package? The package will include space — laboratory and office — and money for staff, equipment, and reagents. Each of these is drawn from a separate pot, so you need to evaluate each of them separately.

Space

Space is always a limited resource. Junior faculty are often assigned a minimal amount of space — ranging on average from, say, 400 to around 1000 square feet. The research techniques you will use and the equipment you need will determine how much space — the absolute minimum — that you have to have. For modern molecular/cellular biology research, a well-laid-out 400-square-foot lab can hold 2–3 full-time staff and some equipment, but will be tight if some of the equipment is large. Some 600 square feet can house 3–4 staffers and hold a reasonable amount of equipment.

Though it really does depend on your particular needs, 1000 square feet is a large lab. A lab of this size can fully support at least 5–6 full-time staff members and lots of large equipment. It takes at least one if not multiple grants to financially support 5–6 full-time staff. So, as a rule of thumb, every lab employee needs about 150–200 square feet of lab space, equipment included. But it's only a rule of thumb.

So, depending on what kind of work you intend to do, a large lab may not be essential. Still, you need to be sure that you have the space you need as you expand your resources and your scientific family. In contrast to single-investigator laboratories, the recently popular large, multiple-investigator joint labs have built-in mechanisms to grow and contract with funding and rank (see "Space Odyssey" by Lila Guterman in the 10 December 2004 issue of the *Chronicle of Higher Education*, subscription required). If the space offered is not sufficient to allow you to succeed, then you must justify why you need more: the equipment won't fit, or a separate room is needed for your ebola virus work, surgery, PCR, and so on.

The condition of the space — and the layout — is also important. Sometimes, the space may need to be renovated to your needs. A careful renovation can help mitigate a space issue, but renovations are expensive and may eat into your package. If the space needs renovating and your start-up funds have to pay for it, be sure there's enough money in the start-up package to get it done.

If you choose to negotiate over space, this will count as your first important negotiating point. If you win what you want, that's one important victory. Then again, if the space is unacceptable, and no amount of renovation can save it — or if the money for renovation isn't there — then this position requires no more consideration. Negotiations over!

Funds

The next most important point of negotiation is the start-up funding package. The good news is that start-up packages have risen dramatically over the last 10 years, mostly due to competition among research-oriented institutions trying to hire the most promising young faculty. Depending on how your prospective chair likes to do business, your start-up package may be offered as a lump sum or as a tally of individual items such as equipment, staff, reagents, and so on. Regardless of whether you are faced with lump-sum negotiations or individual categories, you will have to add up all your anticipated costs and be prepared to negotiate for more if the sum offered is insufficient. Yes: this means you need to have a very clear idea of what you need — and how much it costs — before entering the negotiations.

As you think about negotiating start-up funds, it is important to remember that these funds are to provide you with what you need to initiate your career, not to maintain it. You will be expected to pay for all your research costs from your extramural grants as soon as possible; still, you should make sure the start-up money is sufficient to pay all your expenses until you've had a chance to establish another revenue stream.

Equipment: Equipment is a special category because it is a one-time cost, and the department/institute owns it. Some of it may even be shared by other scientists. Because she knows the ropes and your work, an experienced chair is likely to have a pretty good idea of how much the equipment you need will cost. Very likely, the search committee discussed it with the chair before making you an offer.

You need to be aware of the costs, too. So walk around your current lab and tally up the costs for the items that you need all the time, those that you can share (if they are readily available), and those you need only every once in a while. The first two categories, added together, will probably run about $100–150K for a molecular/cellular lab. For scientists working in other fields, the number will range from lower to quite a bit higher. You may be able to win yourself some wiggle room by including in this tally equipment that is already available in your prospective department that you are willing to share. Because they don't have to know this, it allows you to concede other items.

Some very expensive equipment may require a separate negotiation, because it is out of the range of what most departments are able offer as part of a start-up package. In some cases, funds for such expensive equipment items, especially equipment not in the department or building, can be obtained separately from a special pot of resources.

While funds for expensive equipment may be easier to negotiate for at large research universities or research medical centers, smaller less research-intensive institutions may qualify for grant programs that can provide extraordinary resources. Regardless of the source, the chair may have to negotiate something with her dean. If you ask for and receive a special dispensation for expensive equipment, then this may be win number two.

Staff: Because staff costs are recurring, they are often the most difficult to negotiate. Depending on the size of the institution, it is expected that junior faculty will establish a laboratory with as many as 2–3 full-time members during the first three years. This group usually will consist of a technician and one to two graduate students. A mid-level tech and one graduate student will cost $65–75K per year (depending on salaries, fringe benefits, and stipends). Tuition costs, if you are required to pay them, can increase the amount you need by another $20–30K per year.

The negotiating points here are for additional staff — like a very talented postdoc — and the number of years of support. Because recurring funds come from different fiscal-year budgets, you may be able to negotiate for an additional person in year two. Your chair, however, may expect you to have extramural funds to pay for them by year 2. Nonetheless, if you want a larger group, you may want to consider this topic as one of your main points.

Reagents: The costs of reagents are always difficult to predict. One way is to assume that each full-time staff member spends about $15–25K per year depending on his or her need for animals and high-cost specialty reagents such as antibodies and radioactivity. Don't forget to count yourself as full-time staff. It is expected and wise for you to conduct your own experiments and not place your career in the hands of others until you are established. It's easy to justify the above numbers, but if you need more than $25K per person per year, and the budget offered is insufficient, then you will need to state why.

Length of time: The number of years of support for your fledgling group becomes critical as federal budgets become tighter. Because your first grant application may not be funded on the first or second submission, and because you will need preliminary data to support your aims, several years of support is required. Negotiating for three years of support for a tech and a student with supplies is common, but some institutions/departments just can't afford to offer staff salary several year's out. This point may be worth pushing hard on, for win number three or four.

Sometimes start-up monies are linked to a fiscal year or a special account. While you may not care about the source, there may be a time limit on the funds, or the chair may impose one if she believes that the monies should be spent to jump-start your lab. Knowing the answer to the question will help you decide if you need to negotiate this point. You will also need to know what happens if you get funded before your start-up runs out. Will the department cut off your unspent start-up funds if you don't "need" the money anymore after your grant starts? It would be a shame to lose the money if you were brilliantly successful and got your grant within the first few months. This is not a difficult point of negotiation but one that should be explained in the start-up package.

Teaching Responsibilities

You need to know how you feel about teaching when you are considering an offer. Almost all junior faculty are expected to contribute to the teaching mission of the department (see our earlier article "To Teach or Not to Teach"). The questions are how much, and when. If the job is in an undergraduate department, you can expect to teach a minimum of a course/year, more likely one each semester, possibly more. If you are also expected to establish a strong research lab, then you need to delay the full teaching load for as long as possible.

Most new faculty can get a 1-year delay in their teaching load. If you would like a longer delay, put this closer to the top of your list of points worth fighting for. Medical school departments team-teach most of their courses, so the teaching burden is usually greatly reduced. However, it may require a lot more work on your part to teach a lecture about fungal infections to second year medical students if you are not sure what causes athlete's foot and you were assigned the task because you work with *S. cerevisiae*. Regardless of the department, it is important that your teaching responsibilities are clearly spelled out and that you are satisfied with the arrangement.

Service Responsibilities

Committee work. Advising students. Administering graduate programs. Running a core facility. Seeing patients. Service to your department and institution comes in many forms. Some of these responsibilities, such as seeing patients, may be directly linked to your salary, or to the ability of the department to pay your salary. If you are seeking such a position, then this is a critical point of negotiation. For example, physician scientists have the difficult task of balancing their research time with their clinical time. If they are spending 50–60% of their week seeing patients, it will be nearly impossible for them to start up a successful research lab in a reasonable period of time. Conversely, if they do not see patients for an extended period, they will not be very good physicians. Many physician scientists try to work out a split of their time so that they are spending 70–80% of it on research (or 1–1.5 days a week in the clinic).

If you are running a service core of some kind that is unrelated to your research, then a similar split may be appropriate for you as well. Naturally, this split time is a point of negotiation. Additionally, you may also negotiate a delay in the time before you carry out your full service obligations. Of course, if the department is hiring you to fill a specific service obligation, this will not be negotiable, but you will know this from the start.

Salary

Everyone wants the biggest salary. Salary funding is complicated by the fact that it is derived from university funds, research grants, and sometimes from clinical service. Some of the parameters that govern salary include type of institution and department (medical school, undergraduate college); your degree (PhD, MD); cost of living in the region; and what others at your rank and experience earn at this institution as well as "peer" institutions; among others. Even though you and your salary are a recurring cost to the department, there is room to negotiate here, too. You will need to know the basis of what others earn to negotiate this point.

If it is a medical school, the AAMC publishes a book each year — *Report on Medical School Faculty Salaries* — comparing salaries in all the types of schools, departments, and faculty ranks. This very thorough salary survey is a good resource for researching the salary ranges in American medical schools. *Science* also has conducted surveys and reports. Professional societies in other fields also publish salary surveys. You will also know from your other interviews and offers what the current range is.

With this in mind, unless you live beyond your means — or the salary offered is too low to support your family — salary should not have a direct impact on your ability to succeed. The take-home message here is to negotiate for what you need and what is a fair wage for the department, position, and region of the country. All this info should help you make a reasonable counteroffer for salary if you deem it to be a point worth fighting for.

Along with salary is the issue of salary recovery (see *Giving It 110%*). How much of your salary do you have to pay from your research grants, and when will this begin? If you are expected to garner extramural research funds, then it is likely that you will be expected to eventually pay some percentage, if not 50% or more, of your salary from grants or other outside resources. This practice varies from field to field, and seems to be less common in the physical sciences.

While there is probably a department standard for salary recovery, it should not apply to you right away because you will be starting out as a junior faculty member. You may be able to negotiate a delay as to when salary recovery will start or even negotiate the expected amount that you will recover by some specified point. Alternatively, you may be able to negotiate additional funds for your lab if you can recover the department's standard sooner than a specific time. This would be the case if you were able to bring your own grant with you when you start the position.

Get It in Writing

It is important to get your offer in writing, signed by the chair. No exceptions! Make sure the letter says what is included and how long you will have access to the funds. If a lump-sum number is cited, ask to have the categories that you can spend the money on listed; i.e., staff, supplies, equipment, travels, and so on. It may seem a little awkward at first asking your future employer for something in writing, but get over it: Any experienced department chair will respect your efforts to have things well documented.

Summary

Everyone has a different set of points that are important for his or her happiness and success. The above discussion left out many minor points that will come up ... though they may not be minor to you! In preparing for your negotiations, rank your points as: must have, would like to have, and can live without. This will determine which points you argue over and which ones you concede. Remember, this is a business deal, and try to stay focused on the issues you want. It is likely that you will go back and forth a number of times until you have what you need. Be patient. You were the department's choice and the chair also wants you to succeed and will do what she can to give you what she has to offer. In the end you will most likely receive a fair package that will allow you to develop your career.

Good luck and happy negotiating!

The following article on time management issues first appeared on-line in *Science*'s *Next Wave* on April 9, 2004.

Academic Scientists at Work:

WHERE'D MY DAY GO?

9 APRIL 2004

"**5**:30! I hardly got anything done today," your colleague with the curly red hair shouts at you as she passes you in the hall. You think *you* didn't get anything done either, yet there the two of you are standing in the hall yabbering about how the day went by and nothing got done. You both complain that there was no time to finish your experiments, write your test questions, revise your hot manuscripts, meet with your advisees, help with graduate-student recruiting, design the new website for your department, order the food for the department poster session, and pick up the kids from soccer practice at 7:00. But, as you look around, some of your colleagues seem to have it all under control. How do they do it?

The simple answer: organization, time management, and a good tolerance for stress. If you want to accomplish what you need to accomplish each day, you need to take control of your time. This article will discuss aspects of time management and planning that are of particular use to academic scientists.

Tick, Tick, Tick ...

So, where *did* your day go? You started out with the intention of clearing off your desk and responding to the 63 emails that showed up overnight, then getting right to work on your projects. But when you turned on your computer, you saw the front page of *The New York Times*, and decided to check out some of the articles. After 45 minutes of reading the *Times*, you started reviewing the emails and saw an inflammatory missive from one of your colleagues telling you (in **bold letters**) that your GammaZoid collector in the common lab had contaminated the space and to do something about it — **pronto**. You spent the next hour writing and rewriting an equally inflammatory email back to "Dr. Kleen" and another half-hour calming down from the emotional email outburst.

As you went through your stack of papers on your to-do pile, you came across last week's memo from your chair requesting your updated CV. You realized that you've been shuffling this memo around different corners of your desk for a week and that the deadline was yesterday! An hour later, you had your updated — but tardy — CV and you were ready for lunch, despite the fact that you hadn't gotten into the lab all morning.

The rest of the day was eaten up by lunch off campus with your colleagues, which took up way too much time, again checking *The New York Times* on the Web, responding to emails, talking to people who walked in your open office door, and talking on the telephone to your collaborator in Italy about the wording of four sentences in your joint manuscript. Now it's 5:30, and *that* is where the day went!

Following are some ways to begin getting control over your time, starting with the concept of "the forest and the trees."

The Forest

Before you manage your day, you need to think about managing your career. This is the forest view. The overall ("forest") goal is to be a successful scientist and teacher; if you are a junior faculty member, getting promoted with tenure should be on the list, too. To accomplish the first part of this goal, you will need to design, plan, and execute decisive experiments. Executing experiments requires money, but to get money you need to prove to your peer group that you will use the money effectively.

Thus, one of the most important tasks of your career is to write and publish manuscripts. Manuscripts provide the proof that you can do your job and achieve the goals that you set for your science. Then it starts again: With more publications, you will be able to pursue additional research grants to support your scientific ambition. It takes money to do science, and it takes good (published) science to get money. The sum total of all this coordinated effort is scholarship. In research-centric universities, scholarship is by far the most important factor for promotion and tenure. Even at many teaching-oriented schools, scholarship carries a lot of weight.

Speaking of teaching: That second goal — becoming an excellent teacher — may also be important, to you personally and to your career. To be an excellent teacher, you must be able to spend considerable effort in the design of your lectures, assignments, and exams. This takes time. Preparing tomorrow's lecture tonight may get you through the lecture, but it won't allow you to bring novel and creative approaches to the lecture hall. And it surely won't win you the teaching award.

The criteria required to keep your job, get promoted, and receive tenure must always be on your agenda. While this seems obvious, it is easy to get caught up in all the academic service work and lose site of this goal and what it takes to achieve it. It's easy, in other words, to lose sight of the forest because of the trees.

The Trees

Whereas the forest represents the major goals in your academic life, the trees represent the incremental goals that you must complete along the way. Each manuscript that you publish, grant that you submit, course that you teach, or committee that you sit on is a tree. The leaves on the trees are the actual tasks — preparing figures, writing a paper or grant proposal, preparing and delivering each lecture, meeting graduate students, providing academic advice, and so on. Some of these tasks are more important — and more time sensitive — than others. These two characteristics — importance and time sensitivity — are not the same and have to be kept separate. The key to time management is to prioritize.

PRIORITY CATEGORY	IS IT URGENT?	IS IT VITAL?
1	YES	YES
2	NOT SO MUCH	YES
3	YES	NOT SO MUCH
4	NOPE	NOPE

Prioritize, Prioritize, Prioritize!

Do all the important stuff first, right? Sounds easy enough. But what about all those deadlines for unimportant stuff? There are four basic categories that you can use to help prioritize tasks. These categories are partitioned by their importance and timeliness in a manner described by S. R. Covey (*The Seven Habits of Highly Effective People: Powerful Lessons in Personal Change*, New York: Simon and Schuster, 1989).

Tasks with an "urgent" designation are those that have deadlines. The "vital" designation is for items that are important to your career or personal life. The "forest" aspects of your career belong to category 1 because they are important for your career success and also have time deadlines (e.g., the tenure clock). Items in this category include meeting grant proposal deadlines, preparing your tenure review dossier, and so on. Category 2 is for the "trees"; they usually do not have immediate deadlines. Category 2 items enter category 1 as the due date approaches or passes.

For example, preparing lectures for your upcoming classes moves from a category 2 to a category 1 task as the date of class approaches, if you haven't yet finished your lecture slides, handouts, and so on. Preparing and submitting manuscripts are typically category 2 items, but they can move up if you are in competition with another lab. Category 3 items may be things like getting an order in for a reagent so that you can do the experiment this week instead of next or meeting with a salesperson about a new self-cleaning, non-contaminating GammaZoid fusion collector. Calming a lab disagreement would certainly fall on the urgent side, but its overall relation to your success is not high. However, it is important that the conflict be resolved effectively to avoid the incident from becoming a category 1 affair. The last category reflects most items that are not related directly to your success. Organizing some nonscholastic/teaching departmental event should stay in category 4 as long as you do not let the time component get in the way. In general, the goal should be to spend as much of your time as possible on vital tasks and not to let all other matters eat up your day.

Do It Once

To improve your efficiency at handling routine tasks, respond right away to every request that requires a simple action or response. If the list of chores in the request is complex and multifaceted and you are not going to deal with it now, put it on your to-do list.

Splitting Chores

Sometimes tasks have multiple steps that can be accomplished at different times. For example, reviewing grants and papers is a two-step process. Sometimes, it takes longer to read the grant and write the review if it is done in the same day. Often reading a manuscript or grant application in one day, letting it "ferment" overnight, then writing the review the next morning will allow your opinion on the matter to gel, making it easier to concisely and efficiently write your review. Also, it may be easier for your schedule to split the reading and the writing tasks, rather than trying to block out an extended period of time to accomplish both.

Means of Communication

While email is fast and you can do it on your own schedule, it tends to eat up lots of time. Because the email is right in front of us on the computer screen, we tend to respond to it quickly. If you can't resist opening each incoming message, reset the "get my mail" feature of your computer to 1-hour increments. If this doesn't help, retrieve and answer your emails in the morning and late afternoon *only*.

Although it's not as much of a problem as it used to be, we academic scientists often find ourselves in long telephone conversations with colleagues around the world. These are enjoyable but can eat up your day. If you choose to communicate frequently on the phone rather than by email, stick to business as much as possible and stay on the subject.

For example, working through a joint manuscript on the phone with a collaborator can take all day. If you are the primary writer, have your collaborator fax, express-mail, or email comments to you so that you can include them in the manuscript. This way you only have to discuss those changes that you don't like. If you are the one doing the editing and your colleague is coordinating the document, send your comments to your collaborator and make the follow-up call short. Also, remember that the journal copy-editor will alter stylistic changes anyway, so arguing over them early on is pointless.

Internet News
The Internet is without a doubt the greatest new source of information to come along in a very long time, maybe ever. Its news, weather, stock prices, and 24-hour shopping can also be addictive. So pace yourself. If you must read the news, do it only in the early morning or evening. Do not use a news network as your homepage. Bottom line: Stay focused. Spend as little time wandering the Internet as possible.

Lunch and the Hallway
Other things that eat time are the social gatherings between colleagues. The most frequent are hallway/water-cooler/coffee-maker conversations and group lunches. Each of these venues is important in finding out what is going on in your department and what is transpiring at the upper levels of your institution. Collegiality is important in any workplace.

But, although interacting with your colleagues is important and should be a regular event — especially if you are going to talk about teaching and science — these conversations need to have an ending point. Be aware of your time and make sure that you get back to your office or lab bench. Spend some time developing strategies for extricating yourself gracefully from these conversations. Having a fixed schedule for the day, perhaps with large blocks for lab work marked in, will provide a ready excuse for moving on, while also helping you manage your time.

Committees
Scientists who work in academic positions have to do committee work. It's part of the job. To you, a committee may be a complete waste of time; to others, the same committee may be very important. Committee meetings can be efficient or not, depending on how they are run. For a committee to be efficient, it needs to follow an agenda of tasks to be accomplished. If you are asked to attend a committee meeting, ask for the agenda. Read it ahead of time so that you have the opportunity to think about what you want to say.

If the outcome of one of the discussion items is important to you, discuss it with senior faculty ahead of time. Go into the meeting with a plan. If the discussion item is not important to you, do not prolong the discussion by being a devil's advocate. Finally, have an exit time planned. When asked to be on a committee or attend a meeting, let the organizer know ahead of time when you need to leave the meeting. This may prevent the "how about those Braves, can you believe they let Maddux go?" conversations from starting up and delaying the work of the meeting.

Close Your Office Door

If you close your office door, you will get more accomplished. There are two reasons for this. First, a closed door is the equivalent of an invitation *not* to come in. People will come to your closed door and think, "She must be working, and I just wanted to talk baseball anyway," and then they will leave. The second reason is that it cuts down on the external distractions and allows you to concentrate on the work at hand. The exception to this is that you want to be accessible to your lab staff; make sure that you emerge periodically so that they can see you (and vice versa).

You can also make it clear to your lab staff that they should feel free to interrupt you with important work-related questions even when your door is closed. Although closing your office door sounds like the perfect solution to getting some work done, there is one important drawback: This practice may make you seem "unfriendly" to your colleagues. If all your colleagues have their doors "open," then the advice here is to use this secret weapon only when you need to catch up on your work.

Working the Off Shift

You can get a lot accomplished if you can find time to work when no one is around who needs to take advantage of your time, thoughts, and energy. The hours before 10 a.m. and after 6 p.m. are the most quiet times, most weeks, at a university. Even if you work early or late only one day a week, it will make a big difference. Working Sunday evenings has a special advantage in that it allows you to organize your the week just before it starts. But don't forget to leave time for a personal life.

Working at Home

Some people find that if they really need to make a big writing push or need some reading time, staying at home is the only way they can get control of their time. If you are not distracted by the mailman, the dripping faucet, the need to choose the right background music on the CD, the ice cream in the freezer, or the television — and if you don't need to be near the laboratory — then working at home may be right for you. If you get distracted when you are at home, go back to campus and close your office door.

In summary, time management is critical to the success of an academic scientist. You must do lab work, write grants, write papers, teach, and perform academic service — sometimes all in the same day. So prioritize your work tasks according to their urgency and importance to your career and life. Break habits that waste hours during the day. Be aware of where your day goes, implement whatever strategies you can think of to save time, and do not get so caught up in the day-to-day that you lose site of your objectives.

Oh! Time's up — where'd my day go? Gotta go now. Good luck!

This article appeared on *Science*'s *Next Wave* site first on December 12, 2003 and addresses the issues and strategies to approach when responding to a grant critique.

Academic Scientists at Work:

I CAN'T BELIEVE THEY DIDN'T LIKE IT! — GRANT PROPOSALS

12 DECEMBER 2003

"Can you believe this summary statement? They said my grant proposal was too preliminary, not focused, too ambitious, just a descriptive survey, a large fishing expedition, did not have an adequate animal model, did not provide a clear rationale, and was viewed with a low level of enthusiasm. And that's just in the opening summary paragraph! There are four more pages of specific comments. What am I going to do?"

The good news is that everyone has received some of these comments in their grant proposal critiques. The tough part for the scientist is to determine how to respond so that the proposal is in better shape — *much* better shape — when it is resubmitted. This essay will discuss how to evaluate and respond to your review and provide ways to seek the best advice for improving your grant application. We start with the role of the reviewer because your understanding of the reviewer's objectives will provide you with a reference point for your response and resubmission. The discussion that follows uses the National Institutes of Health (NIH) system as the example, as this is the largest source of biomedical research funding. But the general advice applies to the peer-review systems of other federal agencies and those of private foundations.

Grant Proposal Reviewers

While you may be viewing your grant application as the magnum opus of your life's ambitions and plans — for the next 5 years anyway — a reviewer sees it as one of six to 12 other "magnum opii" projects to evaluate. The reviewer is asked to rank the assigned stack of grants from best to least using his or her experience, and score each of the grants as outstanding (1.0) to not satisfactory (5.0). The reviewer's score allows others in the peer-review group (study section) to gauge his or her level of enthusiasm for the project. Because very few of the grants discussed during the study section meeting will be funded in this submission cycle, each reviewer will choose one or two grants to champion. The reviewer uses his or her knowledge of *science* to decide if your grant is novel, is likely to result in new, important information, and can be accomplished with your experimental plan.

Notice that we referred above to the reviewer's knowledge of "science" instead of a specific "field." This is because reviewers are often asked to evaluate grants that are not in their specific field. This means that your grant must be written so that any competent scientist working in your general scientific area can evaluate (favorably!) the application. So if there was a misunderstanding about the significance of your project when it was reviewed the first time, on resubmission you need to do a better job communicating and selling the concept and its aims to your audience. This is really the most challenging aspect of grantsmanship: creating an original idea and presenting it so that others want to see your idea(s) tested.

Reading the Critique

After you calm down, you need to face the prospect of resubmitting your proposal. The first step is to go back and carefully read and reread the critique. After the second or third read-through, divide the critique up into the types of action you need to take to improve the grant. Major categories for organizing your approach include:

1 Rationale for the experiment

2 Preliminary data/proof of principle

3 Experimental approaches and their alternatives

4 Reconfiguring/jettisoning an aim

5 Data interpretation

As you are doing this, you will have ideas on how you want to respond. Jot them down. When this process is complete, call your program director/program officer and seek his or her advice on dealing the review.

Program Officers/Program Directors

NIH program officers, or program directors as they are sometimes called, serve as scientific liaison between the study section that reviews your grant and the institute that will hopefully be funding your work. These people are YOUR representatives, the ones who can push for your grant to get funded. They are there to discuss your grant and to help you improve your chances of getting funded. Ask your program officer if his or her institute has a program in place to help junior faculty prepare their grant applications (and resubmissions) and how you can sign up for this assistance. [Editor's note: also check out our "Toolkit" piece, Getting an NIH R01 Grant.]

Many NIH institutes provide on their websites detailed information to help new investigators understand the grant review process and improve their grant writing skills. The table below lists some of these Web addresses.

Grant Writing Assistance Sites

NIH Office of Extramural Research
        http://grants.nih.gov/grants/grant_tips.htm
        This links to nine of the best NIH websites that provide extensive "Grant Writing Tips"

National Institute for General Medical Science
        http://www.nigms.nih.gov/funding/tips.html
        "Tips for New NIH Grant Applicants"

National Cancer Institute
        http://www.nci.nih.gov/researchfunding/grants/
        Includes access to book-length PDF for "Everything You Wanted to Know about the NCI Grants Process But Were Afraid to Ask"

National Institute of Neurological Disorders and Stroke
> http://ninds.nih.gov/funding/write_grant_doc.htm
> Very thorough "How to ..." guide that includes dealing with an
> unfundable score

National Science Foundation
> http://www.nsf.gov/pubs/2004/nsf042/start.htm
> NSF Grant Proposal Guide

Each institute sends a representative program officer to observe the study section proceedings and take notes about what happens during the review of each grant. If you are lucky, your own program officer was at the meeting when your grant was discussed. He or she may be able to provide some additional comments for you from the discussion of your grant that was not included in the summary statement. He or she may have read through your grant as well and may have some separate suggestions or comments. The key to talking to your program officer is to *listen* to what he or she says and pay close attention to suggestions about how to improve your grant application. One of the topics that may come up is whether your grant was sent to the most appropriate study section.

Changing Study Sections

Certain study sections — including, perhaps, the one your proposal ended up in — will NEVER fund the type of science that you do, regardless of its importance, impact, novelty, or creativity, because your topic is just not what that study section "goes for." If this is the case, then you may want to ask your program officer about moving your proposal to a different review group. Although this change can be difficult to accomplish, it is not impossible.

You may be able to make the case that the original study section's expertise may not coincide with the project/field that you are working in. To prove that point to the program officer you must know the expertise of the panel. If you don't know all the members, scientific interests can be quickly obtained by searching the member's publication history using PubMed. If you do ask to switch study sections, then you need to have another one in mind. You can use the NIH website to get a listing of the goals of each of the study sections, the types of science they evaluate, and their membership rosters. But even if you do decide to seek a change of study sections, take the criticisms of the old study section seriously and do your best to address them.

Seeking Advice From Colleagues

Now that you have analyzed your critique and talked with your program officer, you should have the beginnings of a plan for resubmission. To cement your plan, you will need additional advice from senior colleagues. Ask several colleagues who have study section experience to review your grant. Give them lots of time to read your reviews and the grant. Did we say lots of time? YES! When you meet with them *listen* to what they are saying and avoid being defensive or argumentative. If they provide only grammatical comments, take those, but make it clear that you are also interested in scientific and other substantive comments.

Responding to the Critique

For most grants, a resubmission is allowed some form of rebuttal or introduction. In this section, DO NOT attack the reviewers. This will not help you. Begin your response by summarizing the strengths and weaknesses offered in the reviews. Then describe briefly how you have responded to these comments and, in the cases of major changes, indicate how they resolve issues raised by the reviewers.

After these opening one to three paragraphs, proceed to specifics using one of two strategies. The first strategy is to deal with each of the reviewers' comments separately. Most people either quote directly or paraphrase the reviewers' criticism, and then provide their response. The second method is to deal with the comments by specific aim, again quoting the reviewers' statements and your changes/responses. This may save space if some of the reviewers' concerns are similar. There usually is not enough space to address each concern; so address the major issues and as many minor comments as you can. You must be concise. This is the first section that the reviewer will read before they launch into reading your proposal, and it sets the stage for the rest of your presentation. Be upbeat!

Specific responses to the critique will vary depending on the category of the reviewers' concerns. Any problems the reviewers have with your rationale/significance must be addressed early in the rebuttal and in the body of the grant application with sufficient clarity that the reviewer will be convinced. Remember: Your goal is not to win the debate on facts, but to win the advocacy of reviewers and study section members.

Sometimes a reviewer will really hate one of your specific aims. You can deal with this in a couple of different ways. The first is to jettison the aim altogether, which will be the end of this problem. If the proposal is still strong without this aim, don't replace it; you may replace it with a new aim that they also hate. Sometimes it is best to take one of the other aims — one the reviewers liked — split it in two, and develop each further. This is a good approach if you can provide each of the revised aims with clear rationale and preliminary supporting data. A second approach is to take the best part of the hated aim and interweave it into one of the other aims.

If you now have the key experiments to *fully* justify and support your investigation described in the hated aim, then presenting this information and rationale could make the aim fly. If this is the case, be sure to get several colleagues to read both the reviews and the revised aim to see if it does fly, instead of crashing and burning up your grant proposal.

Often reviewers state that they do not understand why you are taking the approach that you proposed when it could be done another way, which they proceed to describe. If this was an approach you hadn't thought of — and one that is likely to work — you should incorporate these suggestions into your plan, either as a main approach or as an alternative to your methods. If you stick with your original plan as the main approach or you do not incorporate the suggested experiments, be sure to justify your rationale for their omission. Make sure your senior colleagues agree with you.

If you lack preliminary data to prove that your system or model is valid, then you need to generate the data before you resubmit. This may include developing specialty reagents such as antibodies, mice, or proteins. As expensive and labor-intensive as some of these experiments may be, you must get this accomplished or your grant resubmission will not be successful.

Data-interpretation issues can come in many parts of the grant review. Problems arise when the reviewer does not believe the model you are testing or the conclusions that you are drawing from your work or that of others. At the very least, you will need to acknowledge that there are other interpretations and that you will consider these alternatives in evaluating your data. You may have to propose experiments to distinguish between the reviewer's view and yours. If the difference in interpretation arises from your experiments, then you will need to provide additional data to back up your claims. Sometimes this problem can be solved with a literature search; sometimes it will require some bench work.

Consultants and Collaborators

Reviewers may question your expertise on a procedure or an experimental approach. Whereas established lab groups are not often challenged on standard procedures, fledgling groups are. Identifying a qualified collaborator or consultant to provide you with reagents, protocols, and the all-important letter of agreement to assist in the project will aid your case. Even established investigators do this. It removes a criticism and it will probably be cost-free.

This Round or Next?

If you cannot provide the necessary data to address the concerns of the study section by the next deadline, you are better off waiting to resubmit until a future round. But what if your Chair is breathing down your neck for you to get your grant in sooner than you are ready? Before you approach your Chair to discuss your rationale, you should have your plan fully developed and on paper. Lay out a written plan that describes what you have to accomplish before the grant application will fly. Indicate the names of your colleagues who have told you that you need to do A, B, and C, and provide a timeline for getting these tasks accomplished. This should show your Chair that you are taking this seriously and that you are listening to the senior faculty — including him or her.

I Addressed All the Concerns, So My Grant Will Be Funded, Right?

To answer this question, there are several points to discuss. The first is the difference between funded and an improved score. Sometimes, the funding level changes such that a fundable score in one fiscal year is no longer fundable in the current year. In this case, a resubmission with minor changes may boost your grant above the pay line.

The second point reflects whether or not you truly *did* satisfy all the concerns raised by the reviewers, and whether they really had a high level of enthusiasm for the project. If there was a high level of enthusiasm for the project and the issues were addressed, it is likely that the score will improve substantially. However, if there is not a high level of enthusiasm for the project, it is possible that additional concerns will crop up, even if they weren't mentioned during the first review. Although this is unfair in many ways, it speaks to problems with the question(s) being asked or hypotheses being tested. If this occurs, under the advisement of your senior colleagues, you may consider refocusing some of the aims/goals of your proposal. Or, as mentioned above, seek another study section.

Getting an Early Start

As hard as it may be to do, get in your grant application well before the deadline, and make sure the penultimate draft is read by at least one of your senior colleagues.

Although the advice offered here appears to be simple, it is often difficult to perform. Our pride and our attitudes get in the way in handling critiques of our work. The key is to read your critique thoroughly and dispassionately, and respond sensibly to the queries and concerns of the reviewers. Ask your program officer and colleagues for their advice and incorporate these points into your revised application. Stay optimistic and persistent through the process. Ultimately, you will prevail if your science is good and your plan is sound.

Good luck!

This article first appeared on *Science*'s *Next Wave* website on February 13, 2004 and discusses salary recovery issues and choosing a job based on how much of your salary you need to recover from grants.

Academic Scientists at Work:

GIVING IT 110%

13 FEBRUARY 2004

"**W**ith this new grant, 110% of my effort will be covered by research grants. I certainly should get a big raise — after all, what else is the department going to do with the extra money?" you say with exuberance. Your colleague replies, "Congratulations on the new grant, Dr. F. Ort. You really are doing great — this is just what the chair wants." But as you talk with your department administrator, you find that there is a problem with your effort allocation. 110% is too much.

But how can that be? You put in at least 30 hours a week at the lab bench, 10 talking with students and research staff, another 8 preparing and delivering lectures, and at least 6 hours on a variety of committees, journal clubs, and research meetings. You must be putting in 60 to 70 hours a week. Clearly, your full effort is 150% or greater of a 40-hour workweek, so why shouldn't you get paid for it?

The simple answer is you can never be paid for more than 100% of your effort, no matter how much time you put in. But that doesn't mean you can't benefit from your own success; the increased funding and salary recovery may entitle you to some reward. This article will discuss the concepts of salary recovery, what it means to you when you look for a job, and why Dr. Ort's effort distribution may cause some problems. We will also discuss various salary recovery incentive programs.

What Is Salary Recovery?

Universities pay faculty, in part, to teach and disseminate information to their students. But universities also hire faculty to further the institution's academic mission of scholarship, research, and service. It is helpful to think of the university's different missions as different businesses, or at least as different divisions of the same business. The cost accounting of each of these enterprises is distinct, even though the same person is carrying out each of the missions. It's easy to decide who should pay for the time you spend teaching a class, but who should pay for the time you spend in the lab, seeking the next biomolecular advance?

More than 50 years ago, an agreement was reached between the federal government and a group of universities. It was decided that federal extramural funding agencies would provide universities with faculty salary support for the projects that they fund. In this way, universities are able to recover some of the costs they incur in doing research that serves the public — and national — interest.

In turn, faculty are partially released from their university responsibilities and spend this portion of their time and effort on the research projects that are paying a portion of their salary. This system, in which federal agencies such as the National Institutes of Health (NIH) and the National Science Foundation (to name just two) gain the research expertise of university faculty in carrying out their scientific objectives, has led to the preeminence of American research universities and science.

At the extreme, if 100% of one's salary is paid by extramural research grants, then the faculty member should have no university responsibilities. Thus, salary recovery is the amount of one's effort that is recouped from extramural funding sources. Implicit in this definition is the fact that one can only have effort totaling 100%, irrespective of the number of hours worked in a week. At the other extreme, some universities may not allow faculty to buy release time for research during the academic year. This may be the case if faculty are supported on 9-month contracts. Often, these universities allow faculty to focus on their research over the summer months and receive salary support from their extramural funding sources during this time.

Effort Distribution

Given the above definitions, it is clear that Dr. Ort is overcommitted in his research projects and his efforts in his job with the university. How can he fix this? Most funding agencies will allow faculty to redistribute effort as long as they feel that the reduction of effort will not jeopardize the success/completion of the project. For NIH grants, the easiest way to change one's effort is during the submission of noncompeting renewals at the beginning of the grant budget year. NIH will allow principal investigators to alter their percent effort by 25% of the declared effort with written justification on one of the form pages.

That is, if you are at 40% on a grant, you can usually reduce it to 30% for the next year. In all other cases (greater than 25%, mid-fiscal year, or other agencies), you will need to discuss your proposed changes with your program officer(s). Remember: Part of the evaluation of your original grant proposal was related to the amount of effort you said you would devote to the project. Reasonable arguments for reduction in effort can include the following:

- You have accomplished a substantial part of the proposed work.
- You have established the proposed system and it no longer requires the same amount of your time.
- You have hired a whiz-kid postdoc who has made great progress and your supervisory effort can be reduced without jeopardizing the project.
- You have staff members who can carry out your experiments.

Although there are other examples, the rationale for changing your effort must be factual. In viewing these examples you can see that it would be difficult for Dr. Ort to reduce the effort on the new grant, as he cannot make the above arguments. If you attempt to reduce your effort on a new grant, the program director may also reduce your funding associated with that effort. So, be careful what you propose. Importantly, before you take any steps, you should discuss these with your department administrator and chair to make sure that you do it right and that they are on board with your plan.

Although the above discussion seeks to clarify the issues surrounding percent effort on research grants, it does not address the fact that you are spending a lot of time performing university activities during the time that you should be working on research. As a junior faculty member, you know that you must do your part for your department with respect to teaching duties, mentoring, and service obligations if you want to become a tenured member of the department. Yet, you also know that a major part of your tenure decision will be based on your research and scholarship excellence. So, how does one do both at the same time?

Fortunately, the definition of "research time" includes a wide range of activities. Your research time includes the time when you (a) mentor graduate students on research projects; (b) prepare for and teach an advanced course related to your research field; (c) review manuscripts or grants in your field; (d) discuss science with a seminar speaker; (e) work out the design of a new core facility that you will use for your research program, and so on. Other activities, such as preparing and lecturing for an undergraduate course, do not qualify. Interestingly, the time involved in writing a new grant proposal cannot be charged to a current grant, even if it is the competing renewal. Thus, unless your institution has policies that require you to maintain 100% of your effort supported from research grants, you are better off keeping at least 5% to 10% of your time free for all those extra activities. After all, they really don't take that much time, do they?

Salary Recovery and Choosing a Job

The issue of how much of your salary a university will require you to recover on grants is critical to the overall support that your department/institution will give you during your career, and therefore should be near the top of the list of factors when deciding on taking a position. Although there is no official upper limit aside from 100%, most principal investigators request 20 to 40% salary recovery on their grant applications. (An NIH survey of noncompeting applications found a median of 30%.) Thus, if you are required by your department to recover 60% of your effort, you will need to have at least one large grant (with 60% effort) or several smaller grants, each covering a portion of your salary.

Obtaining several grants (each with a small percent effort) is difficult for several reasons. The first is that it is simply difficult as a new investigator to be able to justify to a study section that you have the ability to run more than one project in such a small amount of time. As you gain more experience and demonstrate your productivity as an independent investigator, you prove that you can run your own show. In time this "penalty" disappears, but when you're just starting out, grant reviewers are likely to look askance at a big project with a small (say, 10%) time commitment from the principal investigator. The second reason is that funding agencies require that each project be distinct. This means you have to keep up with multiple literature sources and direct your attention to "non-overlapping" projects. The ability to successfully do this comes with time and experience.

Many faculty find it easier to compete for private funding sources — often disease — or subject-specific philanthropies — where their field of interest may give them an advantage compared to the public funding agencies. And here's another hint: Collaborating with colleagues as a co-investigator on their funded projects can help you recover salary in small bites. If successfully arranged, your time and effort may be able to be easily covered from multiple sources.

One other really important point: It's a very good idea to stagger the end dates of these sources of funding, so that you can keep your funding (and salary recovery) consistent over the course of your career.

Having to recover a large percent of your salary could mean that you will spend all those extra hours in a week (168), trying to obtain — or maintain — two or more grants. Doesn't sound like a great deal, does it? If your institution requires a lower amount of salary recovery (30% to 40%), you may be able to run your program successfully on one grant and not worry so much about your salary. This being said, there are advantages to the first option. The advantages are discussed below.

Why Do Institutions/Departments Require Salary Recovery?

Scientific research is expensive. Top-notch research institutions can increase the number of academic scientists they support by paying only a part of their salary. Increasing the number of faculty can increase the intensity of the research environment, raise the scientific prowess of the institution, and create unique intellectual environments that would not be possible with fewer members. These are the clear advantages. This strategy will also increase the research and F&A ("facilities and administrative") dollars flowing into the institution and hopefully support more research and related activities. The danger is that the university may rely too heavily on extramural revenues to support their faculty. When the funding climate chills and funding levels fall, overextended universities may be hard pressed to support their faculty and programs.

In addition to extramural research grants, faculty salaries are derived from a multitude of university sources. Your department budget has funds allocated to cover all or part of your salary. The greater the portion of your salary that the university can recover from extramural grants, the more funds the department has available for other activities, such as service contracts, secretarial support, seminar programs, or computer support. Additional funds recovered from salary support at the department level reduce the amount of funds that the university needs to commit to that department as well. This financing practice makes it possible to create a variety of incentive programs to encourage faculty to get more grants.

Incentives can include salary supplements. So, Dr. Ort may be able to convince his chair that he should get a salary supplement once he succeeds in fixing his effort allocation. A recent article in *Science* (W. T. Mallon and D. Korn, 23 January, p. 476) discusses the philosophy of supplement programs and warns that they are good when funding is high but may turn sour when funding levels fall.

Incentives can also include unrestricted, discretionary funds for research development. This could be a real plus and allow a faculty member the freedom to pursue an idea that could lead to a new grant. Additionally, such funds could be used to pay for a new postdoc, computer, or fancy gizmo in the lab. Truly unrestricted funds may even be used for such morale-boosting activities as taking the research group out for lunch to celebrate some research accomplishment. However, you need to be sure that this expenditure is legit. If not, you should consider using your own money to reward your group for their performance.

It is important to know the basic effort and salary recovery parameters of your current job. If you are seeking a job, be sure to ask that these issues be laid out clearly. Incentive programs are often touted as a big plus, but it is just as important to know what would happen to your status in the department if you do not meet the minimal salary recovery requirements.

Good luck, and give it your best 110%.

This article appeared on *Science's Next Wave* site first on August 13, 2004 and addresses the issues and strategies to staffing your lab. It was initially entitled "Hirin' and Firin'."

Academic Scientists at Work:

STAFFING YOUR LAB

13 AUGUST 2004

"**W**hat I really need is someone who can take my ideas, design the experiments, perform them with the right controls, and then write the paper," you say to your colleague with the curly red hair. You both agree that's what you need and that you both did all of this as postdocs. Why is it so hard to find the right people? Are they out there? Where are they hiding? And how do you know them when you see them?

The good news is that there are a lot of talented, bright, hardworking people seeking jobs in science. The trick to finding one is to decide exactly what skills and characteristics you are looking for and develop an organized plan for interviewing and selecting the right person. In this article, we will discuss hiring professional staff: research technicians and postdoctoral fellows. We will also discuss what to do when it doesn't work out and you need to make "adjustments" to your laboratory work force.

What Do You Want?

Hiring good people is the challenge of every profession. Companies dedicate whole departments to staffing, which they now call "human resources recruitment." Because academic research labs are in fact small businesses, principal investigators need to develop a sound strategy to ensure that the right person is hired. And the smaller the research group, the more precise the selection has to be.

Start by making a list of attributes the successful candidate should have. For example: Candidates should be smart, motivated, easygoing, conscientious, reliable, careful, and neat. Well, that was easy. But this list describes the perfect candidate in every profession. Actually finding someone with all of these characteristics is rare. So which ones are essential and which ones can you live without? Being smart and motivated are basic requirements for professional staff in science. But the other criteria will vary, depending on the level of the position and whether it is a research tech or postdoc position.

Postdoctoral positions are "training" positions. Postdocs expect to learn something from you and accomplish something tangible — publishable — in your lab so that they can move on to their next postdoc or their first faculty position. Characteristics of ideal postdocs — in addition to being smart and motivated — include the ability to function independently, to bring their ideas into your research program and enhance the outcome in the design and execution of experiments, and the ability to write. If your postdocs cannot carry out any of these duties, you will not be happy with their performance.

Much of this can be determined from their CVs. If a postdoc applicant has no recent publications, there is little reason to believe that he will suddenly manage to collect the data and produce a publishable manuscript. "Past performance predicts future productivity" is a truism that can be applied here. So now we know that the ideal postdoc must be smart, self-motivated, independent, and a competent writer.

Depending on career goals, research technicians fall into two classes: "permanent" (that is, two or more years) and short-term (fewer than 2 years) employees. The short-term tech could be one who has just graduated from college and is looking to gain some research experience before continuing her education. The permanent category includes individuals who love doing science but do not want the responsibilities associated with being a PI. To you, the difference between the two groups will be cost, experience, and time to train. A senior person will cost more but will know how a lab works and can usually come up to speed quickly. A "green" tech will take some time to train and will likely have to be shown how to make a 1-M salt solution. Regardless of which group you choose to recruit from, your tech must be reliable and easy to communicate with.

Selection

Research Technicians

Begin by making sure that your HR department has a clear idea of what you want. These offices sometimes screen out applicants before you see them, and your ideal person may be scratched off the list because you checked a box on the job requisition forms that required the applicant have experience running agarose gels. If the list of people you get doesn't have what you want, ask HR to send you the whole group of applicants for your job opening. Among the green group, education and GPA may be the factors to consider, whereas in the seasoned group work experience counts.

For a tech position, select at least five to 10 people to interview. Yikes! That seems like a lot of people and a big commitment of time for the interview process. It is, but it is important that you see a wide range of qualified candidates so you can make the right hiring decision. Try to interview all the applicants during a 2-week period. This will allow you to comparison shop and also to capture your best prospect before he gets a job someplace else. Schedule 45 to 60 minutes for each interview.

You have two distinct objectives for the interviews. You need to find out all about *them*, but you also want them to think that *you* would be great to work for. So reserve the last 5 to 10 minutes of the interview for discussion of your science.

The Interview:

Issue 1. Were they on time? If not, they might not be reliable. This might sound harsh, but this is hardball. Of course, being on time today doesn't guarantee that this will always be the case.

Issue 2. Are you able to communicate with them? This is crucial to a constructive relationship. Start out by trying to get the candidates to relax, perhaps by engaging in "intro chatter." Thank the candidates for coming and ask non-challenging questions: Were the directions okay? Parking? The weather? How did they find out about the job? Limit the chatter and try to swing into interview mode, or you may spend an hour with the applicants without learning anything about them or their qualifications.

Issue 3. Is the candidate's education and research experience sufficient for your position? Ask questions such as Where did you go to school? Did you do undergraduate research while a student? What type of lab courses did you take? (This last one is only for greenies.) Can you tell me about your research experiences? At this point, see if you can get candidates to open up so that they are talking about the details of their lab experience; you want to know if they understood what they were doing. Ask them about the general problems they were trying to solve in their projects; you want to make sure they understood the rationale behind the work. You also want to find out about the techniques they performed, which techniques they liked, what they hated, and so on.

Questions like these allow you to evaluate the research skill set of each candidate and determine which assays he or she has performed — or watched someone else perform — more than once. It also allows you to determine whether you can understand each other. The important thing here is to get the candidates to talk.

Issue 4. What are their career plans? The answer will help you understand what motivates the candidates. It will also let you know if they plan to attend medical school in the fall, just in case they forgot to mention this on their applications.

Issue 5. If there are hazards in the lab, physical or ethical, make sure to mention them at this stage. Say "This job will require that you work with small amounts of radioactive materials, mice, biological toxins, pathogens, embryonic stem cells, and hazardous chemicals. Are you okay with this?" If a candidate has objections or flinches at words like "radioactivity" or "mice," then your job opening and the applicant are not a match.

Although you cannot legally ask about a candidate's personal life, you should be very clear about working hours and the policies for sick leave, vacation, overtime, and how the two of you will communicate about these and other aspects of working in your lab. Be clear about the working hours and other expectations at the outset so that there is no misunderstanding. There's nothing more important than making sure your new hire knows what is expected.

Issue 6. What do references say about your potential new employee? Read the letters, if there are any, and then call the authors. This is important. Some people will write letters extolling the perhaps-imaginary virtues of just about anyone. However, they are less likely to mislead you on the phone. So, *call*! Make sure the conversation focuses on reliability, independence, work habits, notebook accuracy, data quality, and the ability to work as part of a team. Your goal is to figure out if these former employers would hire them again if they had the opportunity.

Postdoctoral Fellows

It's true; most of the best fellows go to the big lab groups at the big-time institutions. So, how do you get good applicants? You need to get the word out that you and your research program exist. Send letters to people in your field and to graduate programs at other institutions; visit the posters at meetings and engage the presenting graduate students in conversation. If you are giving a talk at another university, have lunch with the senior grad students; interact with them and let them know that you are hiring. In short, sell your stuff and advertise. Ads in journals are worthwhile, but you may not get the same type of applicant as people you have made contact with or people who have "found" you by virtue of their interest in your field and your work.

We've identified the ideal attributes: brilliant, productive, well trained, well read, creative, writes like Emerson, and works all day. How do you determine whether an applicant has these qualities? Productivity and training are apparent, as these are indicated by the applicants' CVs. If you are looking for someone who will bring specific scientific expertise or new technology to your group, these too are easy to spot. To find out the rest, you need to have your best applicants visit your lab group for an interview and present a short seminar. But before you invite them, call their current mentors. Emphasize finding out the applicants' levels of independence, creativity, and ability to solve problems. You also may find out about their motivation and how much of their project was of their own design.

The seminar presentation will allow you to find out what they know and if they can think on their feet. Because you will want the applicant to say "Yes" if you make a job offer, do not rake him or her over the coals during the seminar. Invite some colleagues to the seminar, not just your research group, so that they can give you their opinion of the applicant if you are unsure. If you're inexperienced, the opinion of a respected senior colleague, backed up with a sound justification, will not only help you make a better decision; it'll teach you something, too.

In addition to the seminar, you will need some time to talk one-on-one and get a read on what they can contribute to your program. Presenting and discussing a set of open-ended project ideas will engage a good candidate's imagination and get him or her to offer some suggestions. This is what you are looking for: someone who can think, solve problems, and move your program forward. So pay careful attention to what the applicants are saying. And, by the way, the candidate should already know your work. If a postdoctoral job applicant hasn't bothered to PubMed you and download and read your papers, then either they are not serious scientists or do not know how to approach science. Forget 'em and move on.

The Offer

Now it's time to make decisions. If you've found someone, make your offer. Your institution has probably established tech salary ranges based on years of experience. You should know what the range is before the interviews and be prepared to discuss salary if it comes up. Offer a fair or equitable salary that is within your budget. If the candidate wants more and you are interested enough to negotiate, find out from your department administrator or from HR how far you can take the negotiations and still remain equitable to similar employees at the institution. If your budget allows, you might be allowed to offer the possibility of a small raise after a 6-month evaluation.

Postdoctoral salaries are based on years of experience and the budget that is available. Although the National Institutes of Health National Research Service Award salary guide serves as a useful scale, many investigators only consider the year 0 salary on the scale. Thus, the bottom value at year 0 has become the standard in most places. You can offer more money if you have it and you think the candidate is a star. Alternatively, if you have sufficient discretionary resources, you may consider offering incentives such as a salary supplement if the candidate procures her own funding. Independent postdoctoral funding is a win–win scenario, as your budget may be used to hire an additional person or to add funds to the fellow's project.

If the people you have interviewed are not up to your standards, hold out for someone better. Don't settle for someone you're not happy with; this will result in frustration and failure for both you and your employee. So be patient.

Termination

It's been 2 months since the new tech started, and nothing is getting done. The new tech avoids eye contact, doesn't communicate, and you can't figure out what he's doing with his time. In short, the situation is not productive. But firing someone isn't that easy; you must allow for due process in conveying your dissatisfaction to the new tech.

In general, due-process procedures require that you inform the employee that you are not satisfied with his or her performance in very specific, job/task-related terms and that you allow time for the employee to correct the problem. If you neglect to handle this situation correctly, you may spend the next 2 months dealing with the fact that you didn't do it correctly. Laws and regulations vary from place to place. Contact your department administrator and/or HR department to find out how to proceed. If the employee achieves the goals you set out in writing, which have to be reasonable and consistent with the job level, then you should be happy.

Terminating employees is a hard decision because getting fired is a seminal event that will affect the person in many unpleasant ways. So be very careful in your hiring decisions and be sure that your new lab members get the orientation and training they need to succeed. If you approach the staffing of your group with care, it is likely that you will seldom have to terminate a member of your lab in order to maintain the progress and quality of your research..

Good luck and happy hiring.

This article appeared on *Science*'s *Next Wave* site first on October 8, 2004 and discusses what to do when projects do not seem to be working out and how to avoid dead end research projects.

Academic Scientists at Work:

THE RED HERRING

8 OCTOBER 2004

"**Y**ou won't believe this new data my grad student presented at our lab meeting. It really is amazing; it opens a whole new area of research!" you say to your colleague with the curly red hair. As the weeks go by, a second set of experiments are also very compelling. But there is something fishy about the way the results are coming out. Everything seems nearly to add up to the conclusion you want, but not quite. While the initial experiments repeat successfully, others do not clearly support the conclusions. This goes on for 6 months until you say, "What is going on? Should I let the student continue this project or is it leading him — and my lab — into a black hole?"

Welcome to the world of the scientific red herring: a set of experiments and results that lead you down a path that ultimately offers no useful conclusion. Biological systems are by their very nature complex, and sometimes all the pieces of the puzzle are not accounted for by the experimental paradigm. Other times, an early, provocative experiment is over interpreted or inadequately controlled.

To be a successful scientist and a successful mentor, one needs to know when a system is not working out and when it is time to change direction or punt. This article will discuss how to avoid the red herring — staying with false leads too long — and when to call it quits on a line of investigation that is simply not going anywhere.

Recognizing and Avoiding Dead-End Projects

How strong are the data? When you are doing the experiments yourself, you know how clean the data are and that the protocols are being followed correctly. As you begin to supervise the experiments of others, you will find that the quality of the data can vary widely. If the data are not clear, perhaps the interpretation has more hope than substance.

So what's the solution? Make sure your student's (or tech's, or postdoc's) data are of publication quality. In the ideal world, all experiments would be publication quality; this, indeed, is what we should all aspire to. The adage that we can always do it again is simply inefficient and may lead to wrong interpretations and wasted time. Adopting "Do it right the first time" as a lab motto may save you months on each project, adding up to years of savings over the course of your scientific career. That could make the difference between tenure and no tenure, and between an illustrious career and a pedestrian one.

Protocol drift? As protocols in the laboratory get passed down from one individual to another, changes of procedure may accumulate. Sometimes these changes are for the better, sometimes not. The issue of "protocol drift" is serious because as protocols change — usually because someone in the lab thinks a modification will save time — the parameter being assayed may also change. The larger the research group, the more rapid the drift. In the worst scenario, you won't know that protocol drift has occurred, and huge amounts of time and money will be wasted on wrong conclusions.

Establishing a protocol-alteration policy for your lab is a good idea. One approach: Changes to lab protocols should be proposed at lab meetings for everyone to hear and discuss, and then approved by you.

Do you have a proof-of-principle experiment? Exciting new data and paradigms are alluring, and great fun to follow up with experimentation. But sometimes we — and especially our enthusiastic but inexperienced grad students and postdocs — get caught up in "Wow, what if this is really true?" Good scientific practice with clear hypotheses — including experimental designs with solid positive and negative controls — should eliminate most errors that might send your research group down a wrong path.

But sometimes that single "proof-of-principle" experiment is missing. Such experiments let you know that your assay system, reagents, and/or model system are likely to give you clear interpretations of your data. Prevention point 1, therefore, is to make sure that all initial experimental systems that you pursue have a single set of experiments that prove that the assay system is accurate and faithful and that the reagents that are being used are precisely what you think they are.

Once the assay system and the reagents are accurate, point 2 is to design an experiment that produces unambiguous results and lets you know that the hypothesis can be proven. In point of fact, breaking down your experimental designs into smaller steps with yes/no outcomes will strengthen most of your experimental paradigms and allow you to interpret your data accurately. The sooner the yes/no experiment can be undertaken, the better.

When to Change Direction

The above discussions deal with adhering to fundamental and irrefutable scientific practices, but even when these are followed projects sometimes lead to a dead end. This can be especially true of lead-in projects that are given to incoming students and fellows, where the hope is that they will be able to develop it into a hot paper or dissertation. It is critical to a young investigator's career to know when to rethink and redirect a project, and when to stick with it. Below is a list of questions that may help you to make the decision to change direction or continue with the current plan. Your decision to change direction should be influenced to some degree by your tenure status and productivity. The more productive you have been, the easier it will be to overcome time wasted on projects that fail.

How long have you been working on this project, and how many months will it take to get to the key experiments that address your central hypothesis? The exact answer to this question depends on the quality of the data being produced and the time it takes or should take to do the necessary experiments. Another factor in the time it takes is the level of experience of the person doing the hands-on work; see the next question. If experiments to address the central hypothesis seem always to be in the distant future, then it may be time to redirect. Certainly, six months of serious effort by you or your staff should be more than sufficient time to decide if a project is worth the investment of more time. Of course, as you become more senior and have accumulated success, investing significant time in exploratory goals and out-of-the-box ideas may lead you to the next great breakthrough in your field.

What is the technical skill level of the person working on the project? Over your career, you will find that all students/fellows/techs are not equivalent in their ability to perform experiments, and that while someone will be good at performing one protocol, he may not be capable of another that someone else in the lab finds routine. This describes one of the basic difficulties in supervising students and fellows. It's not just the quality of the project and the student that matters; it's also the fit. You will have to match the student/fellow with the project to be able to achieve the highest level of productivity. If the person is relatively junior, still developing his skill set, and seems to be stymied by a critical step, you may want to assign this step to someone else in the lab or find someone to collaborate with, or outsource the step to a company.

If a senior, highly skilled person does not appear to be making progress, then perhaps it is due to the initial hypothesis or a basic problem with the power of the technology/protocol to answer the question. If this is the case, there are two choices: (1) approach the question from different angles; or (2) scrap the project. If you opt for the first choice, but still have trouble resolving the question, cut your losses and scrap it.

How risky was the initial experimental goal? Most high-risk experimental pursuits will not end up on the cover of *Science*; that, after all, is the very nature of risk. If you started a conversation with your new student with, "Let's see what will happen if we do," then you probably assigned that student a high-risk project. This is a fine approach. But give your student and the project a time limit and make sure that a proof-of-principle experiment is on the short list.

But the Project is Listed in a Grant as a Specific Aim!

The necessity to complete a grant-specific aim when the experiments are not working can sometimes lead to a huge time investment but only a minor publication, or none at all. Grant proposals are just that: proposals, documentation of proposed scientific work. Not every proposed experiment will work as predicted. If you intend to compete for a renewal of your project, you will need to demonstrate progress, and perhaps even demonstrate that the reason that you did not complete a specific aim is that you proved the hypothesis false. Many grants take this into consideration during the annual progress reports, where you can redirect your plans and pursue related questions and hypotheses.

Help from Colleagues

Your colleagues are valuable resources. You can get the most out of them in two ways. First, if you have graduate students, you should make sure that your students have regular committee meetings attended by all of their faculty committee members. There is a national trend to have at least two such meetings per year. If a student's project is not progressing, ask her to schedule a meeting as soon as possible. The second way to tap into your colleagues' wisdom is to invite a few of them — colleagues with expertise in your area — to discuss your project over lunch, your treat. Approach this as a committee meeting for *you*. Providing your colleagues with an outline/aim handout may keep the meeting focused.

Summary

The core advice here is to follow good scientific practices. Plan proof-of-principle and yes/no experiments early in the process to be sure that the main hypothesis is correct and that your goals are attainable. Be conscious of the time (in calendar months) being spent on moving — or failing to move — the project forward. Importantly, if a project is not going well, put a time limit on when a specific goal or proof must be attained. If it is not attained at that date, change direction and move your other science forward.

While it is fun to have high-risk projects going on, most of them fail. It is therefore important to make sure that you have doable and important science always going on in your lab. Don't put all your eggs in one basket unless you're sure it can hold them.

Good luck!

This article appeared on *Science*'s *Next Wave* site first on July 11, 2003 and addresses issues related to where to publish your papers and what it means to have a paper appear in one of the top journals rather than a low-ranked journal.

Academic Scientists at Work:

PUBLISHING AT THE TOP
OF THE HEAP

EMORY UNIVERSITY
11 JULY 2003

"**W**ow, an article in *Big-Time Science Magazine*! Congratulations! You're sure to get tenure now!" So you say to your buddy with the curly red hair as you walk down the hall. During this walk/conversation, the two of you may question why you made this statement, what the real differences are between journals, and why some are considered to be more prestigious than others. This article will discuss these issues and shed some light on the process of the peer-review publication process.

To begin to look at the reasons behind that statement, it is important to consider the functions of different journals. Although all journals strive to publish the best science, the areas that they cover differ. Some are broad in their coverage of scientific disciplines, whereas others deal with a discrete area of science. Still other journals publish work from the members of their scientific society, such as the American Society for Microbiology's *Journal of Bacteriology* or the American Association of Immunologists' *Journal of Immunology*. Additionally, by publishing articles discussing the political and sociological aspects of science, some journals, such as *Science* and *Nature*, serve as information resources on the pulse of all science.

Types of Journals

Journals can be divided into four general classes: broad top-of-the-heap, near-the-top, society-level, and specialty or sub-specialty journals. The broad top-of-the-heap journals have very low acceptance rates and attempt to publish articles that make conclusions that shift the current paradigm in a field. Articles in these journals tend to be highly focused and may not provide in-depth coverage of the area. Like the authors contributing the articles, these journals are also concerned with being the *first* to publish information that may be of interest to their broad audiences. Thus, there is some competition between these journals. It is also true that manuscripts may get rejected from broad top-of-the-heap journals not because the science isn't spectacular but because the area is no longer "hot."

The near-the-top journals typically represent specific fields or general areas of science (e.g., immunology, neurobiology, developmental biology, structural biology, cell biology, or molecular biology). Over the last few years, some of the top-of-the-heap journals have created high-profile spinoff journals that specialize in a particular area. These are, typically, near-the-top journals. These journals attempt to publish cutting-edge manuscripts that tell complete and in-depth stories. They, too, have low acceptance rates.

Society-level journals publish the bulk of the work coming out of most laboratories and are therefore the workhorses of the scientific publishing industry. These journals have moderate acceptance rates and — surprisingly — may have more citations per year than do the top and near-top journals discussed above. This last fact indicates that these are important journals and that the work published in them is of high quality.

Specialty and sub-specialty journals publish work in a restricted area. Although this work may be of high quality, too, the readership may be limited to only those in the field.

The Review Process

It is important to appreciate how manuscripts are processed, and how and when peer review plays a role. Several different systems are used, and each journal has its own system in place. One thing that's important for you, the author, is the quality of the review and the length of time it takes for your work to get reviewed and, once accepted, into print.

The most basic distinction between journal review processes is whether manuscripts are triaged. Triaging processes are used by editors who attempt to prescreen manuscripts so that they send out for review only those that have a chance of being selected for publication. Many of the broad top-of-the-heap and near-the-top journals employ triage systems. An advantage to you as a manuscript submitter is that the triage review may take only a few days, so you really don't waste a lot of time waiting to find out if the manuscript has a chance of being published. Of course, the disadvantage is that the editor may not appreciate the significance of your title and abstract, and your paper will be returned without the input of an expert in your field.

Journals that review every paper before a publication decision is made run their editorial processes in different ways as well. In a single-step, hierarchical organization, an editor-in-chief is responsible for each decision and relies on an editorial board and a host of reviewers to help make accept/modify/reject recommendations. The editor-in-chief's office finds the reviewers and solicits reviews before making a decision.

Other journals use a multiple-level editorial board to spread out the work of handling the large number of manuscripts received each year. In a two-step system, the journal office receives the manuscript and assigns it to a member of the editorial board, who, in turn, either solicits reviews or reviews the manuscript personally. These editorial board members make recommendations directly to the editor-in-chief.

If the journal receives thousands of manuscripts a year, a system such as this one will require a huge editorial board. For this reason, many large journals use a multistep system, with three (or more) levels of editors. One of these editors will solicit reviews and make a recommendation to the editor at the next level up in the chain. This system works best for journals that receive a lot of manuscripts and maintains its efficiency and quality by having a small number of people making the final decisions on the manuscripts.

Although the editorial decision-making processes are different for different journals, perhaps the most critical question from the author's point of view is: *Do reviewers rate a paper differently if it is sent to the top-of-the-heap as opposed to lower-ranked journals?* In general, the answer is yes! The biggest difference, though, is not one of absolute quality, as you might expect. A good experiment is a good experiment, regardless of where it is published. All journals expect this. The difference lies in the reviewer's perception of the novelty and importance of the work for the broad field to which it relates. Thus, similar comments about experimental design and interpretation are often seen in manuscript reviews from different journals, and the same reviewer may accept the work for the society-level journal but not for the top-of-the-heap journal if he or she feels that the work is not novel enough.

Journal Rankings

Going back to the hallway discussion, that comment you made about your buddy being assured of getting tenure was based on your perception that the journal *Big-Time Science* is better than other journals. Your perception may be due to the journal's low acceptance rate and/or the large number of people who would read the article. Together, these reasons might begin to describe your view of the likely scientific impact of your buddy's manuscript. Your view, though, might not be correct.

The impact of an article or journal can be measured directly by the number of times the average article is cited in other articles. This number — the Impact Factor — is a real measurement often used by chairpersons (or tenure dossier reviewers) to measure the prowess of a faculty member coming up for promotion. The assumption here is that articles published in journals with high impact factors count more than do articles published in journals with lower impact factors.

What exactly is an Impact Factor? The Institute for Scientific Information's Web of Knowledge (www.isinet.com) provides such comparisons between journals, and it can tell you how many times a particular article has been cited. For a journal, the Impact Factor is defined as the number of citations a journal receives in a given year for articles it published over the previous 2-year period, divided by the number of articles it published in that period. Although this math may seem fuzzy, the bottom line is that the higher the number, the higher the journal's ranking. See the table for a comparison of journals in the field of immunology.

JOURNAL CITATIONS AND IMPACT FACTORS FOR 2002*

Journal**	2002 Total Citations	Impact Factor	Articles
ANNUAL REVIEWS OF IMMUNOLOGY	13,709	54.5	26
TRENDS IN IMMUNOLOGY	1,331	15.5	64
IMMUNOLOGY TODAY	11,363	12.9	0
CURRENT OPINION IN IMMUNOLOGY	7,895	12.9	96
ADVANCES IN IMMUNOLOGY	3,199	10.5	6
NATURE	326,546	30.4	889
NATURE IMMUNOLOGY	6,297	27.9	134
CELL	139,765	27.3	350
SCIENCE	296,080	26.7	987
GENES & DEVELOPMENT	45,227	18.8	268
IMMUNITY (CELL PRESS)	19,541	17.5	149
J EXPERIMENTAL MEDICINE	65,261	15.8	310
PROCEEDINGS NATIONAL ACADEMY SCIENCES	315,820	10.7	2911
MOLECULAR AND CELLULAR BIOLOGY	66,521	8.8	774
J IMMUNOLOGY	104,794	7.0	1666
J ALLERGY & CLINICAL IMMUNOLOGY	18,947	6.3	294
EUROPEAN J IMMUNOLOGY	21,579	4.8	406
INFECTION & IMMUNITY	39,706	4.0	925
AIDS	14,183	6.0	334
IMMUNOLOGY	8,448	2.7	179
CELLULAR IMMUNOLOGY	4,634	2.0	61

*Data derived from ISI Web of Knowledge (www.isinet.com).
**Listed by Impact Factor.

For the sake of comparison, some top-of-the-heap and near-the-top journals are included. Notice in this example that the society-level *Journal of Immunology* has an impact factor of 7, whereas the near-the-top immunology journal *Immunity* has an impact factor of 17.5. However, the *Journal of Immunology* publishes 11.2 times more papers and has nearly 5.3 times the number of total citations as *Immunity*. Thus, it is likely that some papers in the *Journal of Immunology* have more citations than those in *Immunity* do. Also note that *Science, Nature,* and *Cell* have the highest impact factors in this list, suggesting that these journals are, as you would expect, the top of the heap.

If you continue your search on the Web of Knowledge site, you will find that review journals and review-and-methods articles have the highest impact factor. Surprised? Consider that when you're writing a paper it's easier to reference a whole field of work with a single citation than it is to cite all the relevant primary papers. Review articles, therefore, are usually cited more than primary work is. Although this kind of objective quantification is something that we, as scientists, like, the high ratings of review articles show that these measures can be biased. Importantly, such quantitative measurements do not *evaluate* the science itself, which, of course, ought to be the most important measurement.

Publishing in High-Quality Journals

Do the journals I publish in really matter when it comes to my tenure decision? Your tenure decision will be based on an evaluation of your scholarship, teaching, and service to your institution. Depending on where you work, scholarship may be the predominant (or even the only) area considered. If this is the case, you will be judged on the impact your publications have on your field and the promise that you show for future productivity. Thus, you will need to publish your work in strong, highly visible journals that are read by members of your broad field of interest. Remember that in making your promotion/tenure decision, your department will seek the advice of leaders in your field and the broader area of your research. These outside reviewers will — let's hope — have seen your papers as they came out, and they will have followed your career through the literature. Your publications are more likely to be noticed if they are published in quality journals, so publishing in society-level, near-the-top, and broad top-of-the-heap journals will help your cause.

This doesn't mean that publishing your work in specialty or sub-specialty journals is bad. It's a very rare lab that produces only science of high potential impact and broad relevance; even if you're focused on important problems, you're bound to produce some science that's just as good but of significance to a narrower audience. Although publishing in specialty or sub-specialty journals may not aid your case as much as papers that are published in higher impact journals will, it does help in terms of promotion, as long as the sum total of your published work influences your field.

The Bottom Line

Almost 20 years ago, an adviser commented to a group of us discussing where to send our papers that it doesn't really matter what journal your work is published in, because if the science is good enough people will find it, read it, and cite it. This statement is even truer today, thanks to the ease with which the literature can be searched and articles can be downloaded, which saves scientists the effort of trudging over to the library stacks.

Some may believe that publication in *Big-Time Science* is equivalent to two or three society-level articles, but the most important thing is to get your work out there where people can see it by publishing regularly in journals that are widely respected, read, and cited by your peers. Make sure your most important work is published at least at the level of your scientific society journals. If your work has more heat and is closer to the cutting edge, you should definitely send it to the journals closer to the top of the heap; after all, you can't publish there if you never send your papers to them in the first place.

Good luck!

This article appeared on *Science*'s *Next Wave* site first on September 12, 2003 and presents strategies to help you respond to critiques of your manuscripts.

Academic Scientists at Work:

I CAN'T BELIEVE THEY DIDN'T LIKE IT! — MANUSCRIPTS

12 SEPTEMBER 2003

"What do they mean, it's 'too ambitious'? How do they think I am going solve the Big Problem if I'm not ambitious? Who are these idiots? They aren't qualified to review my work! None of the experiments they say we need to do will tell us anything! They are just giving us busy work!"

Whew! Wasn't it a relief to get that out of your system? If you have made statements like these, you know that within a few minutes, or days, you will collect your thoughts and decide on a reasonable response to that negative grant review or rejected manuscript.

Almost everyone in science has received a nonfundable grant score or had a paper rejected — usually both. If this is your first experience with such a rejection, congratulations and welcome to the club! The key now is to know how to move forward, to understand the appropriate roles of the reviewer and the reviewed, and to determine how to respond to critiques so that you get that grant funded, eventually, or that paper accepted. In Part 1 of this series we will discuss the ins and outs of dealing with manuscript peer review. Part 2 will focus on responding to grant critiques.

The Job of the Reviewer

The reviewer is just fulfilling a role. It's not personal. And although you may think that you know the name of the person who reviewed your work — you may even be sure of it — you're probably wrong. Journal editors tell us that when authors blame Dr. Stukittume for a negative review, Dr. Stukittume usually wasn't even on the list of potential reviewers. Besides, even if Dr. Stukittume did in fact review your paper, it doesn't matter. The need is the same, whoever the reviewer might be: to move the work forward and figure out how to satisfy the reviewer.

The job of the reviewer is to determine if the work is suitable for publication in the journal to which it is submitted. As discussed in "Publishing at the Top of the Heap," different journals have different criteria that the work has to meet. Reviewers must justify their opinions on acceptance, revision, or rejection of each manuscript. Reviewers justify their recommendations to you through their critique of the work.

However, what you see and what the editor sees are two different things. As you most likely are aware, as a reviewer you get to provide a numerical score or qualifying statement ranking the significance and novelty of the work, the quality of the data, and so on. Reviewers also recommend to editors whether the work should be accepted, returned for revisions, or rejected. But almost all journals have a section where reviewers can make "top-secret" comments to the editor about what they *really* think about the work. Because you don't get to see these comments, you must infer what you must do from the editor's letter and the anonymous reviews. Sometimes it's easy. Sometimes it's not.

The Job of the Reviewed

That's you. Your job is simple: to get the work published. The reviewer of your manuscript expects that you will at the very least address his or her comments. This is very important, since most revisions are re-reviewed by the same people, so they will be looking for you to acknowledge and consider their comments. Reviewers' comments address a range of categories, including novelty, significance, and relevance; the quality and novelty of the experimental design; data interpretation; and style and presentation of the data. The approaches we recommend for responding to each of these categories are discussed below.

Assessing Comments on a Rejected Manuscript

If your manuscript was rejected, the first question to ask is "Why?" Consider the categories listed above. Rejections based on novelty or significance and relevance to a field indicate that the paper was submitted to the wrong journal. To correct this, reassess your work and choose a more appropriate journal. If the paper originally went to a top-of-the-heap journal that publishes only work of broad significance, then consider sending the manuscript to a journal closer to your field, such as the journal of your scientific society. But if the work was rejected for these reasons from a journal that represents your field, you will need to point out the work's significance and how your work adds to what has already been published before you send it out again. However, if you cannot point out how your work adds to your field, then perhaps you will need to wait on publication until you results do in fact add to your field.

Some top-of-the-heap or near-the-top journals reject all manuscripts they do not immediately accept — that is, there aren't any conditional acceptances — and the letters they send out do not explain this fact. If you need clarification on what a rejection letter really means, call the editorial office and ask. If you still can't figure out what to do, seek the advice of a senior colleague in your department. Your colleague may be able to suggest some options that did not occur to you.

Before you press the reformat button and hit the print keys to produce another version of the manuscript for a different journal, you may want to consider the rest of the comments. You may think it unlikely that Dr. Stukittume — the one who *really* stuck it to you, not the one you thought it was — will get your manuscript again if you send it to a different journal. But the world of science is small, and even if you're changing journals, you aren't changing disciplines, so there's a good chance you might get the same reviewer. Some of those comments may improve your work and increase the likelihood that it is accepted. All in all, responding to those comments would be prudent, even if you send the paper elsewhere.

Reviewers almost always comment on the experimental design and the quality of the data. While everyone has a different way of doing an experiment, in the end the data and method of data collection must support the conclusions that you draw. When reading the comments and looking through your paper, ask yourself these questions:

- Is the title of the manuscript supported by the data?
- Do the data support your interpretation?

- Is your interpretation the only interpretation the data support?
- What controls are necessary to nail the point you're making?
- Is there a better way to collect the data?
- Are the results statistically significant?
- Can you get a better autorad reading, etc., than the one that you submitted?

If the reviewer comments on any of these issues, you will need to address them either by doing additional experiments or by providing more information, discussion, or justification before resubmitting your manuscript.

Assessing Comments for a Major Revision

Many journals provide authors with a second or, sometimes, a third chance to get their work accepted. This is most likely when the work is sound and interesting to the reviewer but is incomplete in either experimentation or interpretation. The letter you receive will likely state that the work is not acceptable in its current form. The letter may also state that if you wish to submit a revised manuscript after more work has been done, you will need to indicate how you responded to the reviewer's concerns. A journal editor's assumption is that you will follow the reviewer's advice whenever possible, even if you aren't happy about it.

Responding to experimental/data issues requires work — sometimes lots of it — and work takes time. To minimize the amount of time and effort, you must prioritize: What are the most important points that the reviewer wants addressed experimentally? Start doing those experiments right away. But sometimes the experiment being recommended simply can't be done in your system. This is not the end for this paper; you may be able to perform a different experiment that would support the conclusion just as well. Often reviewers and editors are happy with this "bait-and-switch" tactic. If a requested experiment is easy to do, just do it, even if it doesn't tell you anything you did not already know. This sends a message to the reviewer and editor that you are doing your best to follow their advice.

Assessing Comments in a Minor Revision

Sometimes you will receive a letter that tells you that the work is "accepted upon satisfactory responses to the reviewer's concerns." Congratulations! You are almost there. Don't mess it up now! The concerns here usually can be addressed by doing simple experimentation, acknowledging the reviewer's interpretation in the manuscript, or adjusting a few words here and there. These are all easy to fix, so do it. You can actually say: "The possibility exists that the system may also include the brilliant interpretation of reviewer 74; however, much of the data presented here and also by Superstar et al. argues that the system will behave as predicted." It might be best to leave out the "brilliant" part, though; the reviewer may think you're being sarcastic. If a reviewer suggests that you change some of the wording to make it more palatable, do it. In the end, the copy-editor may change the wording to something completely different, anyway.

Responding to Critiques

Your letter to the editor should start politely. Response letters should state that the author thanks the reviewers for their time and effort and their contributions to the work. Moreover, almost all say that addressing the comments of the reviewers and/or doing the recommended experiments strengthened the work. This is basically true, and it tells the editor that you paid attention. Of course, if you decided to do nothing, do not say that you followed the advice of the reviewers. Instead, you should provide a point-by-point response to each reviewer's concerns. If your response is supported by the literature, quoting papers and supplying references will strengthen your point. In places where you and the reviewer agree, you should note in the manuscript where you have made revisions reflecting the reviewer's concerns. This will help the editor and the reviewer (if the manuscript is sent out for re-review) locate your changes and determine if you have really addressed the issues. These point-by-point letters are often very long, sometimes longer than the article itself. Be as succinct as you can be while also being clear, and avoid derogatory remarks about the review.

Rebutting a Decision

Yes, you can rebut a rejection decision: You write the editor a letter explaining why you believe the reviewers came to the wrong decision. Note how the above sentence is phrased. It doesn't blame the reviewers for not doing their job. If the rejection was based on a misinterpretation of the results by the reviewer, or the lack of an experiment for which you have the results but did not include, then you may have a shot at getting the manuscript re-reviewed. There is the chance that Dr. Stukittume may see your work again, so saying that the reviewer is an idiot and missed the point won't help your cause. You may, however, say, "Reviewer 2 did not realize that the results said *blah, blah, blah,* and therefore we have now reworded the section to make it more clear."

Before you rush to rebut your rejection, realize that there is a good chance that rebuttal will be rebutted. Your best bet may be to make the changes and submit your manuscript to another journal.

Data Presentation and Poor Writing

Sometimes reviewers have trouble with the way a figure or table is assembled or presented. Certainly, if a reviewer comments that data are not presented clearly, you should fix it. However, to avoid such statements, show the figures to colleagues and ask for their suggestions before you submit your paper.

Receiving a comment that the work needs to be edited by someone who speaks and writes English as his or her primary language is one that should be taken seriously. There are professional science writers who can help. Use them.

Bottom Line

As hard as it is to receive a rejection letter, the key to success in science is to receive criticism as openly as possible and without bitterness, and to respond by incorporating or debating the critique in your revised manuscript. It isn't personal. And remember: You can always revise the work and resubmit it to another journal.

Good luck, and happy publishing.

This article appeared on *Science*'s *Next Wave* site first on **May 9, 2003** and discusses why one should contribute to the teaching mission of their school.

Academic Scientists at Work:

TO TEACH OR NOT TO TEACH?

9 MAY 2003

"Concentrate on your research. If your research is good, no one will care if you can teach. After all, when was the last time someone got tenure for being a good teacher?" Every new assistant professor has heard this more than once. Is it really sound advice? To answer this question, it is important first to fully understand the advice being provided.

The recommendation to emphasize research attempts to impress upon you, the young scientist, the fact that establishing yourself in the world of academic science will be difficult. The mentor (if that is the source of this advice) is telling you that you must: (a) set up a lab; (b) get grant money, (c) get your research program running; and (d) make it successful (i.e., publish, publish, publish). If the school where you are working has a tenure clock (that is, a probationary period before a permanent position is offered), then there is a sense of urgency. You have only a few years to establish a research program. Any distraction — even participating in teaching — could keep it from happening.

The intent of the statement that no one ever gets tenure for being a "good teacher" is to drive home the point about the importance of your research. However, your senior colleague used the word "good" — not "great" — in describing the teacher's level of competency. If you had the potential to be a great teacher, and your senior colleague knew it, then maybe the advice might have been different.

Why Teach?

How can teaching in a research-oriented university hinder your career? All of the possible responses could be summed up in one word: *time*. Teaching in all forms takes time. Consider that, for each class, a professor must prepare a lecture or discussion; assemble handouts; prepare overheads or PowerPoint slides; deliver the lecture; write and grade exam questions; and maybe even have office hours to answer students' questions. Just writing that out was exhausting! New lectures can take between 6 and 12 hours to prepare, even if you are an expert in the field. If you are responsible for a major portion of a course, you may find that you are devoting 20 hours a week to your teaching duties. Even the second time you teach a class you will need to spend time preparing, adjusting the syllabus to reflect changes in the science, or responding to students' concerns from the previous year.

At a minimum, a moderate to large teaching load will keep you from pursuing multiple lines of scientific investigation. In the worst case, it may prevent you from doing your research at all until your classes are over for the year.

All this really sounds like you shouldn't teach, but there are many reasons you should think about your teaching and at least *try* to be good at it. Consider this question: Why do university faculty members teach?

There are several possible answers:

    a. It's part of the deal.

    b. That's why they call them schools.

    c. It's fun and you will learn a great deal about the subject.

    d. Teaching is (or may be) required for promotion.

You may think this list should include "(e) all of the above," you may have chuckled at (b) and (a), and you may think that choice (c) is just our opinion, with which you may or may not agree. But hopefully you also took notice of choice (d).

While most research-oriented universities have faculty members who make comments like the one at the beginning of the article, senior faculty members almost certainly expect that *someday*, after your research program is off the ground, you will contribute to the educational mission of the department. It is, indeed, part of the deal. This "someday" often begins with limited teaching assignments in the second or third year of most faculty appointments. In accordance with the importance of teaching, most universities include teaching as one of the three main areas for promotion and tenure evaluation. So, although it's a good idea to focus on your research at first, by the time you come up for tenure you'll be expected to be carrying your weight as a teacher.

Being a Team Player

With the case made that teaching is part of the deal of being a professor and will be required for promotion, there are several other issues to consider as you engage in the teaching responsibilities of your department. Start with the positive aspects of teaching. Teaching integrates you into your school. You are now part of the enterprise. You will meet colleagues associated with your class and colleagues who use the room after you, especially if your class regularly runs over time.

You will also meet students. Depending on the level of class you are teaching, your audience may include prospective students for your lab. This would certainly be the case if the group consists of first-year graduate students. Teaching graduate students gives you an opportunity to show off your brilliant, bubblin', witty, and, of course charmin' personality. All young faculty members should try to participate in first-year graduate courses, since this is the time when most students make their choices about what to study and with whom.

When should you begin to participate in teaching? Most new faculty members get a 1-year hiatus from teaching so that they can get their labs organized, hire personnel, and submit their first grant applications. If you are expected to get your own research funds (and who isn't?), your highest priority should be to get your grant applications in on time and in good shape for funding. So, if you have a choice on teaching dates, try to arrange it so that your teaching is not within 2 weeks of your grant deadline.

While jumping into the pool at the deep end is a great way to get used to the water, this may not be the best approach to developing a set of teaching skills. Start slowly. If your department is structured so that you can pick up a few lectures a year, this would be the best way to approach the task. If you know in advance what you will be teaching, it is a good idea to sit in on the same classes taught by experienced teachers and get an idea of what you will have to do. After all, observing the great Dr. Mia Goudateecha give a lecture will help you organize your class when your time comes.

You may be required to develop a new course. Upper-level graduate courses are the easiest to develop as they can be centered on a research topic and often include a series of papers for you and the students to discuss. They are also the easiest way for you to review the literature; you will be surprised at what you will learn. If you do opt to do one of these classes, be wary of the number of papers that you assign for each class. Students can and will only read two to three papers for each class. If you assign more, they won't be read and the class will be a drag because some students won't be able to participate. If you have to create a lecture course, choose a good book, and follow it. The better the book, the easier your job will be. Following the outline in the book is sure to make your job, and that of the students, easier and more successful.

Teaching and Promotion

A review of the tenure policies available on the websites of research-oriented institutions, such as Duke, Stanford, Harvard, the University of Chicago, the University of Virginia, and the University of North Carolina, Chapel Hill, indicates that demonstrated success in classroom teaching is a required component of the faculty review process. The factors most often used to evaluate successful teaching include student evaluations and anecdotal reviews from senior faculty. Being designated an "outstanding teacher" and receiving teaching awards certainly helps to enhance the teaching-portfolio portion of a promotion package.

In addition to classroom teaching, success in the training of graduate students and postdocs may also be evaluated. This one-on-one training is time- and effort-intensive. The evaluation of such training is usually made by assessing the outcome. Did the student/fellow do the following:

- Publish high-quality manuscripts?
- Complete his or her degree or training in an appropriate length of time?
- Get a good job after he or she left your lab?

The success of trainees is credited to the mentor as the trainees forge their own careers in science. Having students and fellows who develop into productive and successful scientists is almost always positively commented upon in promotion decisions. (If promotion to the tenured ranks must occur within the first 7 years, then this issue may be moot, because the outcomes for a junior faculty member's trainees may not be known by that time.)

As an acknowledgment of the importance of university teaching and the value of faculty teaching skills, many research universities have established "centers for teaching and learning." These centers sponsor workshops, seminars, and brown-bag lunch events, and provide a library of resources to help both beginning and advanced faculty members learn and improve their teaching techniques. Check out the resources at your institution and take advantage of the help that such a resource center can provide.

Getting Credit

Make sure that you get both feedback and credit for your efforts. As mentioned above, feedback comes in the form of student and faculty evaluations. For student evaluations, set up your own questionnaire (don't rely on the standard one the university provides) so that the responses will be objective and actionable; that is, you want to be able to use the feedback to improve the course. It is pointless to have the students rate whether they think you know the material. How do they know? Rather, have them comment on the organization of the course, whether they wanted more specific examples or more general material, and so on. If your school/department doesn't already have a program to provide feedback to its junior faculty, consider asking a senior faculty member to sit in on one or more of your lectures and provide you with tips and comments. This faculty member will then be able to provide a letter for your promotion folder when the time comes. Moreover, his or her comments will help you hone your skills.

It is important to make sure that your chair is aware that you are teaching and how *much* you are teaching. And just as you should be advising your chair about your research progress, make a point of keeping him or her abreast of your teaching prowess. Last, keep detailed records of your effort. If you teach two lectures in Quantitative Micromacros, be sure to get macro credit by listing the lectures and a brief statement of the lecture content in your ever-growing teaching portfolio. Update this portfolio after each lecture. This will ensure that 6 years after your first lecture, you (and your evaluators) will know that you actually did teach in that course and that you were an active team player in your institution.

Bottom Line

So, what is the answer to the question, "To teach or not to teach?"? For the junior faculty member, the answer is "*Teach. Participate in the teaching mission of your department, but do so in moderation!*" Aside from the fact that teaching may be required for your continued employment, you will benefit from the opportunities to work with your faculty colleagues as well as from the stimulating interactions with graduate and professional students. On top of these benefits, you may even learn something.

This article appeared on *Science*'s *Next Wave* site first on April 8, 2005 and provides an editor's view on the manuscript review process and how to critique manuscripts for journals.

Academic Scientists at Work: On Being a Manuscript Reviewer:

TO ACCEPT OR REJECT?

8 APRIL 2005

"What do you think I should do about this paper?" you ask your colleague. "There are some obvious problems that are fixable, but it is written so poorly, it's going to take me 3 days to figure out how to tell them what they need to do." Your colleague with the curly red hair leans back in his chair. "You think *you've* got problems?" he says. "I have a set of grants that I need to read for study section, and it's taken me all day just to get through one guy's preliminary data. Let's go get a beer!"

Reviewing papers and grants is a time-consuming and sometimes frustrating task. What is expected? How do you decide whether to accept or reject a paper or to recommend funding of a grant proposal? Why should you even volunteer to take on such a difficult task when there are papers and grant proposals of your own to be written?

Good questions. We've got answers in this article, part one of a two-part series. This part addresses the "why" question, provides strategies to help you make those crucial decisions on acceptance/rejection for the manuscripts you are asked to review, and offers tips on improving your efficiency at reviewing. A subsequent article will address reviewing grants.

Why Participate in the Review Process?

Peer review is the dynamic process that allows science to progress with the appropriate checks and balances. In order to support great science, we need a system of objective review by scientific experts. The alternative to the peer-review system would be a closed network of self-referent researchers whose work would not serve the long-term advancement of science.

Moreover, there are direct and indirect benefits to scientists who participate as peer reviewers. First, reviewing manuscripts will keep you up to date on developments in your field. Manuscript reviewers have the opportunity to read about the work in their field months before the work is published. Second, reviewers have the opportunity to support and advance the work of their field in peer-reviewed journals. Finally, the fact that you were asked and participated in providing "expert" review on a manuscript is an acknowledgment that you have established a national/international reputation in your field. Your career advancement, whether you are in academic science, government or in private business, depends on such recognition.

How to Review Manuscripts

All journal editors want to publish the best and most timely work — the "hottest" work — that falls within the scope of their journals. It's also important that the work they publish be well performed, controlled, and that the conclusions are supported by the data.

Reviewers are also asked to comment on whether a work fits the scope or is appropriate for the journal in question.

To provide such a recommendation, you must read the manuscript — obviously — and closely examine the data it presents. But when you are acting as a reviewer, you aren't just reading a paper to absorb the information as you might in reading an already-published journal article. You are making judgments about rigor and quality that scientists reading published articles rely on.

Almost all journals have guidelines for reviewers either on their website or accompanying the manuscript. If you are a first time reviewer for the journal, you should read the guide to see what the editors want you to focus on.

Typically, a reviewer is requested to respond to the following questions:

Major Questions

- Is the work novel?
- Is this an important contribution to the field?
- Does the work invoke a new paradigm or is it simply an incremental advance?
- Do the experiments answer the questions being addressed or test the hypotheses proposed?
- Do the data and analysis justify the conclusions drawn?
- Are the appropriate controls in place to justify the interpretations offered?
- Is the work within the scope of the journal or better suited for another journal?

Minor Issues

- Is the work appropriately referenced?
- Is the presentation brief and clear?
- Can the manuscript be shortened?
- Are alternative interpretations considered?

Now that you have read the paper and answered these questions you will need to decide if you think that the paper is (1) acceptable with no changes; (2) should be returned for revisions; or (3) should be rejected outright. To accept "as is," the experimental design needs to be sound and the data need to be clean and accurate enough to support the conclusions drawn. However, your job is not done here. Higher-ranked journals want you, the expert, to tell them how important and novel the work is, how much it contributes to advancing the field, and whether the conclusions shift the current status — the "dogma" — of the field. "Top-of-the-heap" journals (see "Publishing at the Top of the Heap," July 11, 2003) like *Science* want their published papers to be at the very top of the field but also to be of general interest to all their subscribers. Society-level journals want many of the above characteristics in their papers too, but it may not be necessary for the work to shift the current paradigms.

Assuming the work falls within the journal's scope, manuscripts should be returned for revisions if the work is of importance and a limited number of additional experiments or controls can be added to tighten up the work or provide a truly novel spin. Difficulties arise when some of the work is solid and some of it is not. In such a case, the referee needs to determine whether the work supports the broad conclusions the authors draw. If some of the conclusions are not fully supported, you may suggest an approach/experiment to strengthen the conclusions. Other times, the work may be very good but not novel — others may have already published much of it, or its equivalent.

In the latter case, because important work needs to be validated and researchers get scooped all the time, the reviewer needs to consider two things in making a decision: (1) how much time has passed since that other paper presenting (some of) the same results; and (2) how much new information is added in the "we-did-it-too" manuscript you are reviewing. If the time lag is relatively short (months, not years) and there is new information that expands the observations, then you should consider giving the authors the benefit of the doubt. If the timeline moves to years and there is not a lot of new information, then it may be time to reject.

An outright rejection would also be warranted if the conclusions are not supported by the data. This could be because the work was not performed well and the data are not convincing, or the controls that ensure that the system is working are missing. Other causes for rejection could be if the work provides only an incremental advance and is therefore not appropriate to the *Big-Time* journal where it was submitted. Manuscripts that are purely descriptive, do not address a scientific question or hypothesis, or just describe a new reagent or technique may be more appropriate for a sub-specialty journal than a mainstream society-level journal. Such works have no place in top-of-the-heap journals.

Critiques

In general, there are two types of evaluations: a numerical/qualifier evaluation, where you are asked to check a box or choose a descriptor rating the paper in a variety of categories; and a written critique. The descriptors are very important and are used by the editors to determine the overall ranking of a manuscript. For example, as a reviewer you may be asked to rank the "originality" of a manuscript on a scale of 1 to 5 (1 being the best). If you choose 2, you have put the manuscript between the 21st and 40th percentile. Not too bad, but for a highly ranked journal such a score would probably end up with a rejection.

Comments to the Editor: Most journals have a place for the referees to provide "top-secret" comments to the editors. Here, you can state how you really "feel" about the manuscript. If you are on the fence on whether to recommend accept/revision/reject, briefly tell the editor why. Championing a manuscript for acceptance, especially without revision or in a top-of-the-heap-journal, is not as easy as simply writing: "This is good stuff and should be published in this journal." You will need to briefly state *why* it's good stuff in your comments to the editor. "Whys" could include the novelty of the work, its importance to science, or its likelihood of shifting a paradigm.

Comments to the Authors: Now that you have decided what you are going to recommend, it is time to write an informative critique. Do not spend your time correcting spelling errors, misplaced commas, or grammar, because most journals pay people to do this once the paper has been accepted. Instead, focus on what is novel about the manuscript, the degree to which the paper's conclusions are supported by the data, and which parts you believe are not supported. If additional experiments are needed to convince you that the conclusions are correct, *briefly* outline the experiments. If the authors left some critical information out, mislabeled a graph/figure, did not use appropriate statistics to analyze their data, or forgot a key reference, then point this out. Be specific, so that the authors can revise their paper accordingly. Even if it doesn't come back to the same journal (as it probably won't if you or another reviewer recommend rejection), you might be asked to review it for another journal, and you will want to see that they made an attempt to improve the quality of the work. If you like the science but the manuscript is poorly written, tell the authors that they need to improve the clarity of the writing.

Sometimes in our eagerness to be critical, we forget about the people who will read the critique. Avoid emotionally charged or gratuitous criticisms. Remarks like "The data presented in figure 8 are meaningless," can be replaced with "The quality of the experimental data in figure 8 needs to be improved before the interpretation of the authors can be accepted."

Revised Manuscripts

If you have recommended that the manuscript be returned for revisions, you are likely to see it again. Usually, reviewers expect the authors of a revised manuscript to address the concerns raised in the critique. Written critiques for revised manuscripts are typically very short. For papers in which the authors have thoroughly addressed the previous concerns of the reviewer, the written critique can say just that. The paper is now ready for acceptance. That was easy!

Just because the authors were allowed to resubmit a revised manuscript doesn't mean that the paper has to be accepted on this round or any subsequent round. The manuscript must be sound and fulfill the journal's requirements as stated in their guidelines. Thus, if you really want to see the inclusion of another control in the experimental design, or if you still don't believe the conclusions based on the data, you will need to say so in your comments to the editor and the authors. Occasionally, new issues emerge due to either the inclusion of new data in the revision or because you or another reviewer missed a critical point. This being said, it is not fair to the authors for you to come up with a long list of new concerns that were not raised in your previous reviews.

How Much Time Should You Allocate to Manuscript Review?

Reading and writing a critique of a manuscript takes several hours. If you have to do some literature searches or pull some additional references in order to fully understand the work, this will increase the time you have to spend. Journals give you between 7 and 21 days to review a manuscript and submit your review. You don't want to get a reputation for accepting the assignment and then not submitting your review in a timely manner. So, if you know you can't possibly get the review done in time, tell the journal when you think you can do it, or decline the invitation.

You also don't want to provide a mediocre, unhelpful, or biased review. The editors in some journals rank *the reviews themselves* by the quality of their scholarship. This way, they know who the best reviewers are in a field. Some journals also use this data to consider whether someone may be appropriate to add to the editorial board. While such positions usually come as you mature in your career, editorial board positions are a sign of national recognition that will enhance your portfolio and benefit your development as a scientist.

Summary

The manuscript-review process is the cornerstone to scientific inquiry and results reporting. As a citizen of the scientific community, it is important that you participate in this process and that you take your responsibility seriously. By focusing on the questions that the editors want to know the answers to, you can save yourself some time. Leave proofreading to the proofreaders and focus on the science. Your reviewing mantra should be "Be fair, clear, and precise," but don't forget to champion manuscripts that you really like.

Good luck, and don't stop the presses!

This article appeared on *Science*'s *Next Wave* site first on May 13, 2005 and discusses how to review grant proposals and present your reviews at the peer-review group meeting.

Academic Scientists at Work: On Being a Grant Reviewer

TO FUND OR NOT TO FUND?

13 MAY 2005

"How am I ever going to get through all of these grants? There must be more than 200 pages here. I didn't think there would be this much reading, and I only have 2 weeks before I have to send my comments in!" you exclaim to your colleague with the curly red hair. "Relax," he replies, "all you have to do is spend a day on each, and you're done. Piece of cake."

Your colleague is right about the time required to review a grant application. Careful planning is crucial if you're going to complete your reviews and still have time to get your own work done. Furthermore, it's a difficult job; because each grant application is distinct in its goals, approaches, and significance, you need a method to distinguish among the grants and rank the "apples and the oranges."

In this article, we will discuss:

1 The time involved in reviewing an application;

2 The characteristics of grant applications that allow you to rank them in relation to one another;

3 What is expected in the written review; and

4 How to prepare for a grant-review session.

At the outset, it is important to consider *why* you should volunteer your time to serve as a grant reviewer. First, it's for the good of science. The complex system of public and private funding for scientific research relies on expert peer reviewers — people like you — to evaluate grant proposals and direct money to the very best and most important scientific and medical projects. The other reasons are less altruistic. Reviewing grants is a terrific way to gain grantsmanship skills. Grant reviewers learn from the mistakes they observe in the grant proposals they review, as well as from the discussion of the shortcomings — and successes — of the proposals during the peer-review meetings. Serving as a reviewer will also provide you with a crash course on each scientific topic proposed, a benefit to seasoned investigators who serve as reviewers. Lastly, you will get to experience what really happens at a grant-review panel meeting. This experience will be invaluable to you as you prepare your own proposals.

Reviewing Grant Applications

Now that you are convinced that you should participate in grant reviewing, you need to decide if you have the time. The first thing to determine is whether you are being asked to serve as an ad hoc member for just one set of grants at one peer-review meeting, or whether you are being invited to become a standing member on a review team. As a standing member, you will be expected to participate in the process several times a year.

Over the course of your career, this could cost you months of time. If you are a new faculty member, this may not be the best use of your time when you are trying to get your lab going.

On the other hand, being an ad hoc reviewer for a national agency or even for an institutional funding program can be a good opportunity for a new investigator to gain experience in an amount of time that most young scientists can spare while establishing their research programs.

How much time does it really take to review an application?

Depending on the funding agency, you may be recruited to review more than one grant at a time. Sometimes reviews require only a written critique; other times you may be asked to participate in a conference call, a site visit, or a panel-review meeting. Additionally, the page length of the application and what the funding agency wants you to critique differ among the types of grants.

You need to know all these particulars to estimate how much time you will need to commit to the reviews, but here are some rough guidelines. A seasoned reviewer needs a minimum of 2.5 to 3 hours of uninterrupted time to read a 25-page, single-spaced NIH application. If the application requires that you do literature searches and find additional background material, the time increases dramatically. Some grants take all day to read. Many people read a grant on one day and write the critique the next morning. It typically takes about 2 hours to write the critique. You should therefore expect to spend at least one full workday on each grant application you have been assigned.

Typically, reviewers receive their applications 4 to 6 weeks before the review or meeting date. As soon as you get your proposals, look at them. Waiting until the 11th hour to begin the process is a bad idea. If you procrastinate, you will end up miserable, so give yourself plenty of time. That's why they sent them to you a month ahead of time.

Hey, I know that guy!

Avoiding conflicts of interest with grant applications is important. All funding agencies have rules and guidelines specifying what constitutes a conflict of interest. These include co-authorships, training relationships, and financial relationships, to name a few. If you have a conflict with any of the investigators or applications, be sure to let the agency know right away. Scientific conflicts with competitors must be evaluated on a case-by-case basis. If you feel that reading a competitor's application will lead you to change your scientific goals, then you should decline; otherwise, you should review the application, which may be the reason that your expertise was sought in the first place. We all review our competitors' applications and papers. Establishing the ethical line and treatment of this confidential information is part of the career-development process.

Whoa — it's not my field!!!

All reviewers occasionally get stuck having to review an application that is out of their fields of expertise. If you feel that you cannot possibly review a grant, contact the program officer who assigned it to you as soon as possible. Here's another reason to look at your assignments as soon as they arrive. It may help your cause if you see a grant in the group that you weren't assigned that you could trade for. If you can't get out of it, set aside an additional day and be prepared to read some long review articles.

What Should You Look for in an Application?

Along with your stack of applications, each funding agency will provide criteria by which the grants should be judged and ranked. For research grants, many of the criteria are the same. Ranked in importance, research grant applications typically have seven important parameters: (1) significance, (2) novelty, (3) experimental design, (4) preliminary results/proof of principle, (5) progress (if the application is a renewal), (6) the presentation, and (7) the qualifications of the applicant.

Judging the strengths and weaknesses of each of these parameters allows you to rank each of the grant applications you are reviewing relative to the others. It is important to note that we use the word "rank" rather than "fund"; reviewers do not fund grants, they are just asked to rank them. Funding agencies may provide you with a guess at their paylines: the cutoff between proposals that are funded and those that aren't. You're there to sort the grants by priority so that they can fund the best grants first and then work their way down the rank-ordered list.

Using these seven parameters, a "great" grant application has the following characteristics: It addresses a very important problem; introduces novel concepts and tools; has clear hypotheses and uses a clear, straightforward experimental approach; provides clean, useful preliminary data that prove each principle being tested; is written clearly; and is from a person fully qualified to do the work, working at an institution that has the resources to fulfill the requirements of the project. That was easy, right?

So much for the ideal grant; what about those that are not perfect?

Significance

The first point to consider is the importance of the problem being addressed. If the problem is at the top of your list of important questions and issues that this field of science should address, then it makes sense to cut the principal investigator (PI)/applicant some slack if the proposal has a few nonfatal shortcomings. If you think the problem is pedestrian, then there needs to be something else in the proposal to increase your enthusiasm and keep it off the bottom of your rank-ordered list. Although this sounds easy, most applications fall in the middle of the "importance" scale, requiring often difficult apples-and-oranges comparisons. The other parameters gain in importance as the significance wanes.

Novelty

Proposals that introduce and test truly novel concepts should be greeted with enthusiasm — *if* the applicant justifies the new concept and proposes a way of testing his or her new ideas. Because we tend to be rooted in the dogma of our fields, outside-the-box proposals often do not receive the high rankings they deserve. The key to deciding whether novelty is truly a virtue is preliminary data.

Preliminary Data

Depending on the type of grant, the quality — and sometimes the quantity — of preliminary data is crucial to your evaluation. Any novel tool or reagent should be well characterized, with adequate controls. If a grant aims to test a new principle, then a proof-of-principle experiment must be provided. The work doesn't have to be in final publication form — that's why they call it preliminary data — but it should be clear to you that the system is working.

Proof of principle is particularly important when some of the aims are dependent on each other for success. For example, if Aim 1 seeks to create and characterize a new tool, which will then be used in Aims 2 and 3, the ability to create the tool and use it for the intended purposes must be demonstrated. Preliminary data can also serve other essential purposes. If an investigator is using a technique for the first time, preliminary data can be used to demonstrate his or her mastery. If the approach proposed is questionable, preliminary data can demonstrate that the approach is sound, or at least promising.

Preliminary data provide proof that the investigator's ideas are justifiable and doable. Lack of key data could legitimately diminish your enthusiasm for an aim of the proposal.

Competing Renewal Applications

Competing renewal applications are a special category. These applications represent a continuation of a research project that may have very broad and long-ranging goals. Because the investigator has had time to perform all or part of a previously funded set of aims, progress on the previous application needs to be evaluated. In the ideal case, the investigator has completed all the aims of the previous application and published the results in, say, a dozen papers over the last 5 years, with several of those appearing in top-of-the-heap journals like *Science, Nature,* or *Cell* (see our *Next Wave* article, ***Publishing at the Top of the Heap***). This ideal case would be considered outstanding progress.

In contrast, a worst-case scenario would be that the aims were not completed and no papers were published, or, if there were published papers from the work, they appeared in sub-specialty-level journals and had little impact on the initial goals of the previous application. Obviously, this is poor productivity, and your confidence in the ability of the investigator to do the work proposed should be diminished.

What about the 80% of renewal applications that are neither exemplary nor obviously deficient? What is considered solid productivity? As a rule of thumb, one or two papers a year in society-level journals — or better — is considered solid productivity. You must then ask yourself, what if the investigator has fewer papers but has one in a top-of-the-heap journal and covers one or two aims of the grant? This is where journal ranking and perception of scientific reputation come into play. Papers in higher-impact journals are presumed to undergo greater scrutiny, and the work is assumed to represent a major advance in a field. So, extra credit can be given to the investigator for articles in top-of-the-heap journals. Each application will be different, and you will have to judge whether the investigators have done the best job possible with the project they proposed previously.

Experimental Design and Approach

The experimental approach should make sense and stay focused. Although it is often useful for this section to be organized by hypotheses, this style isn't required; some PIs like to pose questions instead. Regardless of the style, the investigator should make it clear what she is going to do and convince you that she has the right controls in place to interpret the data. The investigator should also say what she will do if things don't come out the way she plans. If you can't follow the logic of an experiment, or if the obvious alternatives are lacking, then it is reasonable for your enthusiasm for this aim or subaim to diminish.

Many research plans are designed to test more than one idea and hypothesis. You should judge the significance/rationale, approach, interpretations, and alternatives for each aim separately. You may also have to consider whether the completion of one aim depends on the completion of another. (This is why a "fishing trip" for a new factor or gene at the beginning of a project typically leads to a poor ranking; if the factor isn't identified, the grant can't be completed successfully.) In the end, it is important to determine whether the aims of the proposal are synergistic or antagonistic.

Qualifications of the Applicant

Is the applicant qualified to do the work? First-time applicants, junior faculty, and senior investigators are evaluated differently. In all cases, the publication record of the PI is the key to making this determination. For first-time investigators, the following questions are important: Do they have first-author papers? Many, or only a few? Are their papers in the same field as the field of the application? Were the technologies used in the new proposal also used in earlier papers?

First-time applicants' postdoctoral work is used to evaluate their probability of being able to perform the proposed work. The degree of independence of first-time applicants is also considered. Is he true faculty or a senior postdoc in a senior investigator's lab? Although your enthusiasm for the latter applicants should not be dismissed, you need to be convinced that they will function independently of their bosses and begin to establish their careers. Their resources — allocated space and access to necessary equipment — should be defined clearly.

For scientists who have been independent for 2 to 5 years, the ability to publish on their own is a key criterion. Even if the publication numbers are low, a faculty member who has been in a position for 3 years or so should have a record of publication without his former mentor, and he should be first or last author on some peer-reviewed papers. If the junior faculty member already has a grant, has he published on the topic? This is key to determining whether he can function independently, and whether he can handle more than one research program.

For senior investigators, qualifications may be an issue if the application is in a new field, or if the investigator has not been productive for several years.

Although the productivity issue is difficult for everyone to overcome without new publications, gaining expertise in a new field is not. The simplest way to build credibility for a proposal in a new area is for the applicant to enlist the help of collaborators with strong and published expertise in the new field or technique. Collaborators listed as co-investigators, who will receive part of their salaries from the grant, are best because you know that the PI will expect them to work. Issue covered! Collaborators listed with support letters need to be evaluated based on their letters, published expertise, and what they will be contributing. All collaborators should detail their intents and roles in their letters of collaboration.

Are the required resources readily available? Most of the time this is not a problem, although this can be an issue for new investigators who are still in their mentors' spaces. Regardless of the PI's position, all PIs should have their own spaces and should have access to all the necessary equipment and core facilities.

Presentation

Grant applications do not have to read as if they were written by Hemingway, but they need to be clear, logical, and free of glaring science errors. Although brownie points are not officially awarded for elegance or the overall style, a sloppy application suggests to many reviewers that the investigator is also sloppy with his or her experiments. Still, don't forget that it's the science you're reviewing, not the writing.

Writing a Review

Now that you have studied the application and have developed an opinion about all of the parameters described above for each of the application's aims, it's time to write your critique. Keep in mind, as you read what follows, that at this stage you have already evaluated the proposal's strengths and weaknesses; you have already decided which proposals you like most and which ones you like least.

The style, format, and specific content of the review will depend on the funding agency, but written critiques typically serve three purposes. The first is to convince the panel that the applications you like best deserve to be funded and that the ones you like least don't. Second, the critiques provide the funding agency with documentation — and the applicant with feedback — as to why the grant received the ranking that it did. Third, written reviews provide you with a synopsis of the key points of the application, its highlights, and your concerns when it is your turn to present the review to your peers at the reviewers' meeting.

In preparing your review, start by outlining the major categories you are asked to comment on. Some of these categories are similar to the parameters discussed above, but each agency has its own particular approach. Avoid spending time rewriting the grant for the applicant. And, although you should provide clear guidance, avoid the temptation to describe all of the experiments you recommend in detail.

After an initial assessment of the significance of the proposed science, most of your critique will address the experimental design. If your enthusiasm for a proposal is very high, it makes sense to emphasize the proposal's strengths and not its weaknesses. Your written critique should be scholarly and measured, but the goal of the written review is to bring others — other committee members and those who will make the ultimate funding decisions — around to your point of view.

However, if the proposal has significant strengths but has aspects that temper your enthusiasm, you will need to indicate the strengths but concentrate on the major weaknesses that you think need to be addressed for the project to be successful. This is important because the applicant will use your written critique to address the concerns when it comes to time for resubmission.

Writing a good critique is difficult. It takes several hours to write a 2- or 3-page review.

The Reviewers' Meeting

The peer-review process is at its best when reviewers are asked to justify their ranking of an application to their colleagues on a panel of experts. The process can be intimidating for first-timers. The best way to ensure a good experience is to be prepared.

The first step in preparing is to rank your grants in order of your enthusiasm. This is essential because no one will take you seriously if you rate all the proposals you reviewed as "outstanding." Once the proposals are ranked, choose your top two grants to champion. Not all the proposals you like will succeed, so you need to make sure that the one(s) that you think are the best get your full support at the meeting.

To prepare your presentation, make a bulleted outline of your review, emphasizing the points that you want to discuss. To refresh your memory, read the abstract and your reviews the night before. This sounds more and more like preparing for qualifying exams, doesn't it?

In the presentation itself, stress the significance of the work, the novelty of the system, what the applicants hope to learn or solve, a clever experiment or two, and why you think this applicant is the best person — or team — to do the work. The last item is easily addressed by evidence of past productivity.

For high-ranking applications, concentrate on strengths and do not dwell on weaknesses. Review panel members will use your presentation to rank the grant being presented with the ones they themselves have reviewed. For your low-ranking applications, justify your poor rankings by explaining the problems with the application. Your verbal report should take less than 10 minutes. If you are the second or third reviewer to present and you have nothing new to add to the presentation, just say so, especially if the proposal is not one that you are championing. If it is The One you've decided to champion, make sure the other panelists are aware of your enthusiasm for the proposal, even if you have nothing of substance to add to the previous reviews. Consider repeating something if you think that it will enhance the enthusiasm of the review team for the application.

Occasionally, reviewers will be split in their opinions. This is never good news for the application. How you deal with a situation like this will depend on how strongly you feel about the application. If you have ranked it on the high side and are convinced of its great value, take a stand. Stress the work's importance and argue that the investigator will be able to address the weaknesses raised by the other reviewer. If you have ranked it on the low side and feel certain that the work cannot be done, you will need to explain why, in scientific terms.

Whichever side you're on, always listen carefully to the other reviewers' arguments and be willing to bend if they are convincing. If you don't feel strongly about an application, don't hold up the meeting. Tell the others that you ranked it somewhere in the middle, raise your score if necessary, and let the meeting move along.

Summary

Participating in the grant-review process is a thankless job but one that is valuable not only for you but for the whole scientific community. Give yourself plenty of time to review the applications and write the critiques. Be as fair as you can while remembering that you are not responsible for what is written in the application. Prepare for your presentations ahead of time and be sure to champion the best proposals. In the end, the review experience will increase your scientific expertise and raise your own grant-writing skills.

Good luck and enjoy your role in awarding the grant money!

PART V

APPENDICES
(THE WINNER'S CIRCLE)

The appendices include sample documents, letters, worksheets, spreadsheets, and databases for you to use to help you make decisions and organize your Academic life. Useable blank (larger) copies of these files can be found in the CD that accompanies the book. To use all of the files, you will need to install on your computer Microsoft Word™ and Excel™, as well as Claris FileMaker Pro™. The worksheets are set up as with text boxes in a protected document. This will allow you to use the files without disrupting the formatting. Of course, you could just print the worksheets and use them with pen and pencil too.

While there is considerable variation in the overall format of CVs, they all need to contain the same information. The current example lists where you are, where you have been, and what you have accomplished. Start a file early in your career (like today) and modify it after every event.

Today's Date

Curriculum Vitae
Ima Starr

Address: Department of Immunology
Big Time School of Medicine
Atlanta, Georgia 30322
(555) 555-9999 — telephone
(555) 555-9997 — fax
imastar@immuno.topten.edu

Date and Place of Birth: December 15, 1974
Brooklyn, New York

Family: Single

Education:
     BS 1995; Biology, State University of New York
     PhD 2000; Immunology, Next Tier School of Medicine, Mentor: Dr. T. R. Ainer
     Postdoctoral Fellow: 2000–2004; Big Time University. Department of Biochemistry and Molecular Biology. Mentor: Dr. Jack S. Alright.

Honors and Fellowships:
     —Distinguished Doctoral Dissertation Award 2000
     —NIH Postdoctoral Fellowship Award 2001–2004

Academic Appointments: (■ current)
     ■ **Instructor:** 2004–present, Big Time University, Department of Immunology.

Research Focus:
     Molecular mechanisms of regulation of immune system genes

PUBLICATIONS:
From Graduate Studies with Dr. T. R. Ainer

> Listing papers in this format with your name bolded focuses the reader's attention to your contribution.

1. **Starr, I.**, Besttech, J.B., and Ainer, T.R. 1998. Characterization of interleukin gene promoters in T lymphocytes. *Proceedings of Immunological Journals* 68: 122–141.

2. Readynow, T.M., **Starr, I.**, Besttech, J.B., and Ainer, T.R. 1998. Insightful characterization of interleukin promoters in lymphocytes. *Journal of Important Immunology* 44: 1–13.

3. **Starr, I.**, Pub, D., Besttech, J.B., Readynow, T.M., and Ainer, T.R. 1999. Interleukin promoters respond to signaling in T lymphocytes mediated through the T cell receptor. *Another Big Time Immunology Journal* 88: 1488–1498.

4. **Starr, I.**, and Ainer, T.R. 2000. Nuclear Factor of Activated T cells is activated in our system too. *Journal of Important Immunology* 49: 1279–1288.

From Postdoctoral Studies with Dr. Jack S. Allright

5. **Starr, I.**, Pipetman, R., Emsa, B.E., and Allright, J. 2001. Regulatory factor binding to mutant promoter sequences. *Cell–Type Journal* 15: 772-780.

6. **Starr, I.**, Chuck, U.P., Emsa, B.E., Gell, R., and Allright J. 2002. Control of regulatory factor binding in fetal thymocytes. *Could Be Science* 317:8214–8215.

7. Chuck, U.P., **Starr, I.**, and Allright, J. 2002. Cloning of the big-time factor (BTF-1) in our could be science paper. *Journal of Bigtime Science* 54: 1883–1890.

8. Emsa, B.E., Chuck, U.P., **Starr, I.**, and Allright J.S. 2003. Allelic polymorphisms in BTF-1. *Allelisms* 21: 620–628.

9. **Starr, I.**, Pipetman, R., Vorte, X., and Allright J.S. 2004. Association of BTF-1 with Ohmygaud-7 in primary thymocytes (OG-7). *Cell-Type Journal* 17: 241–250.

As Instructor at Big Time University

10. **Starr, I.**, Vorte, X., and Allright, J.S. 2005. The BTF-1/OG-7 complex is required for Th1 differentiation. *Journal of Bigtime Science* 320: 4320–4323.

> Including manuscripts accepted/in press, submitted or even in preparation is important. Remember that a manuscript is not submitted until it is mailed.

11. Jones, J., Allright, J.S., and **Starr, I.** 2005. Another paper on BTF-1. (Submitted).

12. P'doc, F.D., and **Starr, I.**, 2006. OG-7 is a target of an important oncogene. (In preparation)

RESEARCH SUPPORT

Current Support

> ▪ List all of your support here. Current and previous support is an indication of your productivity and how others reviewed your ideas.
>
> ▪ The style below is similar to the NIH "Other Support" style. However, any style that provides all of this information is fine.
>
> ▪ Type and Grant number, Principal Investigator's name, your role if not P.I., your percent effort.
>
> ▪ List Granting Agency, dates of the award, and total direct costs.
>
> ▪ Title.
>
> ▪ A short description of the goals of the project. You may list more than what is presented here.

1) 1RO1 476   P.I. Ima Starr   60% effort
   NIH NIAID   12/05-11/10 $750,000 total direct costs
   "Characterization of the BTF-1/OG-7 complex"
   This project focuses on characterizing the genes responsible for the broken lymphocyte syndrome an inherited autoimmune disorder that is due to defects in the BTF-1/UG-7 complex.

2) 1RO1 899   P.I. Jack S. Allright
   Ima Starr is a Co-investigator   40% effort
   NIH NIAID   4/02-3/07 $1,250,000 total direct costs
   "Genes involved in autoimmunity"
   This project focuses on cloning the genes responsible for all autoimmune diseases. Genetic and biochemical approaches have been taken to find genes that segregate in disease families.

Past Extramural Support:

▪ NIH, National Research Service Award, P.I. Ima Starr   100% effort
   NIH NIAID   2001-2004
   "Cloning of T cell differentiation factors"

Past Intramural Support:

▪ Big Time University Seed Grant Program, P.I. Ima Starr   60% effort
   Grant number 12   4/04-12/05 $15,000
   "Characterization of the BTF-1/UG-7 complex"

ACTIVITIES AND SERVICE:

Here are some examples of this category. List only the categories that you have participated in. Having nothing under a category won't get you brownie points.

Consulting
> CureAll-Laboratories Inc. 2005.

Professional Societies
> American Association for the Advancement of Science
> American Association of Immunologists

National Service
Peer Review Groups
> Ad Hoc, American Cancer Society; Study Section for Cancer and Immunology. Jan 2005.

Journal Reviewer (ad hoc reviewer)
> Journal of Important Immunology

University Service: (■■ current, ■ past)
> ■ Medical School Postdoctoral Fellows Advisory Committee. Jan 2001–2004.

TEACHING / TRAINING / MENTORING

Students Trained

> These examples should be added in as you get and train students. List Ph.D. students first followed by Master's then Undergraduates. Keeping a running tab on the current positions of your students is important for your future review.

Bill E. Emsa, PhD, — 2004
> Thesis: "Gel shift assays and BTF-1.
> —Postdoctoral: Dr. M. Y. Competitor at Big Time University of the South.
> —Current Position: Assistant Professor, Podunk College, Department of Biochemistry.

James R. Jones, BS, 2001
> Undergraduate Honors Thesis: "Activation of BTF-1 in lymphocytes"
> —Current Position: Graduate School, West Coast Science Center.

Current Students

Doctoral Students
Jean Mann, 2004 to present

Undergraduate Students
Sandy Senior, 2004 to present

Postdoctoral Fellows (Years, Current Position)
- Frank D. P'doc (2004-present)

Graduate Program Memberships
- Department of Immunology Graduate Program

Student Thesis Committees (Year Degree Awarded; Affiliation)

Johnny B. Good	(current: Immunology Graduate Program)
Benny Serve	(current: Immunology Graduate Program)

Course Participation

> This is the hardest area to keep up with. Start documenting your participation early so that you won't forget those hours you spent preparing lecture material. The format below is by year. Some faculty like to organize their list by courses and then list the years, role, lecture hours, and topics. Try to record as much information as possible here.
>
> Year
> Title of course. Level, course number
>   Your role (Lecturer, Moderator, Course Director, etc):
>   lectures and topics

2004
- Concepts in Immunology. Graduate IMM-500.

Lecturer: 1 lecture on BTF genes and proteins.

2005
- Concepts in Immunology. Graduate Course IMM-500.

Lecturer: 1 lecture on BTF genes and proteins.
           1 lecture on OG-7 activation.

Research Presentations:

Invited Speaker:

> Year
> University and Department or Meeting/Session. If session chair,
> indicate here.

2004
West Coast University, Department of Immunology
Lockstone Meeting, Autoimmune Disorders, Session on BTF-like complexes.

Big Time University
School of Medicine
Department of Immunology

(Today's Date)

Search Committee
Department of Macro and Microsciences
Research Utopia University
Big City, GA 30000

Dear Committee:

> In the opening paragraph state what job you are applying for, where you saw the ad or heard about the job, and what is included with the letter.

I am writing to apply for the Tenure-Track Assistant Professor position in your department that was advertised in the latest issue of *Science*. From my review of the research interests of the faculty of your department, I believe that my current and future research focus in T-cell development would complement those of several faculty members. As requested, I have enclosed my CV, a personal statement, and a list of three individuals who could provide you with letters of recommendation.

> In this paragraph, state your current position and previous postdoctoral positions and provide a brief description of your major findings. Also indicate if you have been funded and what your future work will entail. Your description should be exciting and tease the reader to carefully read your CV so that you will be considered for the job.

I am currently an Instructor associated with the Dr. J.S. Allright's laboratory, where I was formerly a postdoctoral fellow, in the Department of Immunology at Big Time University. My research centers on the organization and molecular mechanisms regulating T-cell development. During my postdoctoral studies, I cloned BTF-1, the first of the Big-Time Factors required for differentiation of T cells. Mutations in the BTF-1 gene have been found to be associated with a rare form of autoimmune disease. We have found that BTF-1 forms a large complex with several other proteins, of which we identified one of these as Ohmygaud-7 (OG-7). OG-7 can also interact with an important oncogene, suggesting a potential link between BTF-1 and carcinogenesis. I have recently received NIH funding to continue this work in my own group, where I will be characterizing the BTF-1/OG-7 complex and determining the effect of conditional mutations of these genes in a murine model system.

My long-term goals are to establish a vigorous independent research group and participate in the training of graduate students and postdoctoral fellows. I am also interested in contributing to the teaching mission of your department. I look forward to hearing from you. If you require additional information, please do not hesitate to call (555-555-9999) or email me (imastar@immuno.topten.edu).

Sincerely,

Ima Starr, PhD
Instructor

Appendix 1-3 apears on the ASAW CD, and provides examples of good and bad color combinations and some other tips on PowerPoint presentations.

Institution: Research Utopia Univ.   Date: 8/29/05

Growth Potential	Research Trainees
How many research faculty are currently being recruited: in this department? **2** in other science departments? **Greater than 6**	How many postdoctoral fellows are affiliated with faculty in the department? **25** What percentage have their own funding? **30%**
Is the department planning on expansion of its Space? Yes ☐ No ☒	Are clinical fellows available to join research laboratories in the department? Yes ☐ No ☒
Other: **With the new chair the department had 6 slots, four of which were filled over the last 3 years.**	How many graduate students are working with faculty in the Graduate Program? **34**
	How many students are admitted each year? **6-7**

Faculty Characteristics	
Number of faculty in the department: **16** Junior Faculty **7** Senior Faculty **9** Male/Female **3/4** Male/Female **2/7**	Will you have access to training grant funds for pre-doctoral Yes ☒ No ☐ postdoctoral fellows? Yes ☐ No ☒
How many faculty members have: Funded **25** Non Funded **3** laboratories	Does the department have an MD/Ph.D. Program? Yes ☒ No ☐
Is the Chair of the department a funded investigator? Yes ☒ No ☐	Are technicians typically long-term or short-term? **long term**
How many junior faculty in the department have pursued promotion to tenured rank in the past 5 years? **3** How many succeeded? **2**	Are technicians difficult to find? Yes ☐ No ☒ Other? **Postdocs seem to be in senior labs.**
Are there other faculty in the department who share your research interests? Yes ☒ No ☐ In the Institution? Yes ☒ No ☐	**Research Space**
Do faculty have collaborative projects? Yes ☒ No ☐	What is the size of the typical Assistant Professor's lab? **900 sq ft**
	Is lab space specifically assigned or is it fluid and modular? **assigned**
Are there Program Projects within the department? **one project between 3 faculty** Joint with other departments? no	Do faculty share equipment? Yes ☒ No ☐
Other: **Fun group**	Other: **Common cold rooms, equipment rooms have space for my equipment. Animal facility has space for my mice.**

Core Facilities

Core facilities relevant to your research program? (indicate satisfaction, cost, quality)
Animal facility is run well and costs are OK at 0.50/day/cage
Microchemical facility has state of the art protein sequencing and mass spec instrumentation. They run on a fee/service.
Transgenic facility is new but have made a few mice. Turn around time is slow.
Imaging facility is very good.
Microarray facility is adequate

Overall Evaluation

How do you see <u>your</u> research program developing in 5 years if you were to join this department? Consider the following aspects of your career: National Recognition, Publication Potential, Expansion of Your Program, and Career Development.

My research can do well here if I can recruit students into my lab. The core facilities are adequate.

The faculty are very well funded, each having more than 2 grants. Only three faculty are not funded but they do a lot of the teaching.

The departmental seminar program is great and there is a retreat each year to discuss science

The faculty all seem to get along and the students are very happy.

Rank of this institution's research environment as compared with that of other institutions being considered:

Outstanding ☒   Excellent ☐   I could work here, but.... ☐   Not for me ☐

Other Comments:

Major Plusses	Big Downers
Faculty interactions are plentiful **Good students** **Core facilities are operational** **Shared equipment is new and available** **Faculty are fun**	**Not many postdocs.** **could use more students**

Institution: Research Utopia University   Date:8/29/05

Professional Programs

What professional courses are taught each academic year (Medical, Dental, Nursing, etc.)? **Medical Microbiology**		
How are teaching assignments made for these courses? **The Chair decides**	Are they distributed equally? Yes ☐ No☒	Does funding affect the distribution? Yes ☒ No☐
Are there laboratory courses? **yes** Who teaches them? **A paid instructor with graduate students**		
How is teaching evaluated? **student evaluations and some faculty who come to the lectures**	Is there are Grace Period for new faculty? Yes ☒ No☐ How Long? **1 year**	
Do the senior faculty assist the junior faculty in preparing their lectures? Yes ☐ No☒		

Graduate Programs

How are the graduate programs organized? Departmental Yes ☐ No☒ Interdepartmental Yes ☒ No☐	How many students are admitted each year to the program(s) that you would be joining? 6-7
How do faculty attract students to work with them? **Show 'n tell seminars at the begining of the year**	Do the students perform lab rotations? Yes ☒ No☐ **3-4 rotations during their first year**
What is the typical number of graduate students in the laboratories of:	Asst Prof: **2** / Assoc. Prof: **2-3** / Prof: **2-4**
How are graduate students supported? **first two years by the program and then on research grants**	When do you have to pay them? **3nd year and up**
Is there an institutional training grant supporting the students in these programs? Yes ☒ No☐	Is there an MD/PhD program? Yes ☒ No☐

Teaching in Graduate Programs

What courses are taught? **Introductory and advanced immunology courses**	How are teaching assignments made? **Course director asks faculty to participate**
	Is there a Grace Period for new faculty? Yes ☒ No☐ How Long? **1 year**
How is graduate teaching evaluated? **student evaluations, nothing formal**	What is the average load of thesis committees / Jr. faculty? **seems high but no real number.**

Undergraduate Programs

Is there an undergraduate major in this department? **no**	How many courses do they teach / yr? **n/a**
How are the courses distributed among the faculty? **n/a**	Does funding play a role in this distribution? n/a
Do faculty provide course and career advisement to the undergraduates? Yes ☐ No☐	How many students/faculty?
Are there lab courses? Yes ☐ No☐ Who teaches them?	Is there a Grace Period for new faculty? Yes ☐ No☐ How Long?
Are there Teaching Assistants to help with the courses? Yes ☐ No☐	How is teaching evaluated?

Overall Teaching Responsibilities Evaluation
How does success in teaching count towards promotion and tenure? **It seems to be a minor component, although everyone takes their teaching seriously. Students seem to be with faculty involved in teaching in the first year.**
Rank of this institution's Teaching Responsibilities to that of other institutions being considered:
Perfecto ☐   I can handle this ☒   I don't know ☐   Get me a taxi I'm outta here ☐
Other Comments: **Teaching load will be light, with no more than 10 hours / year.**

Major Plusses	Big Downers
light load **can teach in graduate programs**	**may have to teach mycology lectures to medical students?**

SERVICE WORKSHEET   APPENDIX 1-6

Institution: **Research Utopia University**   Date:8/29/05

Laboratory Service	**Clinical Service**
Is there a laboratory service component to the position, and if so what? **no**	Does the department have clinical service requirements? **not for PhD faculty**
Does this service require your "hands-on" expertise or are you just a supervisor?	What is the ratio of service/research time (i.e., 20% service / 80% research)?
Does the service require consulting time? Yes☐ No☐ How much/week?	Is the service time distributed throughout the year or concentrated during certain months? Yes☐ No☐
	When on service are you expected to be on call (24/7)?
What percent of your time will be devoted to running the service laboratory?	Is a component of your salary dependent on the service? If so, how much? Yes☐ No☐
Does the service benefit your research program? Yes☐ No☐	Is there an enhancement to your salary based on clinical service? Yes☐ No☐ If so, how much?
Other: **Some faculty run clinical Micro labs, but this is not part of my job description**	
	Are you fully prepared for this service assignment? Yes☐ No☐

Teaching laboratory	How are clinical assignments made?
Is there a teaching laboratory associated w/ the position? Yes☒ No☐	
When is the course taught? **spring semester**	Will you share your service practice with other faculty members? Yes☐ No☐ If so, who?
How many classes/week? **3**	
How many students / class? **20**	
Is there permanent staff for lab set up?	How will the service be coordinated?
Are there Teaching assistants involved in the course? **yes**	
How many other faculty members are involved in the lab? **2**	Is there secretarial support for the management of the practice? Yes☐ No☐

Administrative Service	**Other Service Requirements?**
Will you have an administrative position, and if so what are your duties? Yes☒ No☐ **I will be asked to run the seminar program in year 2**	**New faculty are asked to participate in graduate student recruitment.**
Will the service require consulting time? Yes☐ No☒	
Is there secretarial support for the service? Yes☒ No☐ **A secretary will handle all of the travel arrangements**	

Overall Summary of Service Component of Position

Major Plusses:**Light service load. Seminar program will be a bonus in my second year as it will allow me to invite scientists in my field.**

Big Downers:none...

Rate 1-5

1-Terrific ☒	2-Not Bad ☐	3-OK ☐	4-I don't know ☐	5-Lousy ☐

Other Overall Comments:
Service situation is great

JOB COMPARISON WORKSHEET

Institution: *Research Utopia Univ.* **Date:** 8/20/2005

Departmental Chair Dr. M. Musculus **Phone:** 404-555-1234

Appointment - Terms	Start Up - Package
Rank: Assistant Prof. Track: **Tenure**	**Research Funds**
Start Date: **Jan 1 2006**	Total Dollars — **550,000**
Yrs to Tenure: **7**	Equipment funds — **200,000**
Other: **come up in year 6**	Staff funds — **350,000 to be split between staff and supplies**
	Supply funds:
Teaching	Duration: **3 years**
Course Load/Term: **1 course/ year**	Carry Forward Policy — **no more than 40K after 3 years**
Grace Period: **1 year**	Other: **Must purchase equiment in year 1**
Role of Teaching in Tenure: **minor**	**Graduate Student Support policy?**
Teaching Lab? **no**	**Department pays first two years of student stipend.**
Other: **Teaching load is 10 lectures / course Medical School course**	
Service	**Lab Space**
Clinical: **none**	Sq ft: **850** — Number of rooms: **2**
Laboratory: **none**	# benches: **5** — # desks: **5**
Time:	Warm/cold rooms **1** — Yes ☒ No ☐
Grace Period:	Dark Room? — Yes ☒ No ☐
Other:	Specialty rooms — Yes ☐ No ☒
	Renovation required? — Yes ☐ No ☒
Salary Terms	Windows? — Yes ☒ No ☐
Salary: **$80,000** Months/Yr: **12 months**	Other: **Shared common equipment space and tissue culutre space**
% recovery required: **50%**	
By when? **June 1, year 3**	
Other: **Dept will support for one year if no salary support.**	**Office Space**
Fringe Benefits **lots**	Sq ft **100** — Set Up funds: **not separate**
Health, Dental, Vision **Medical includes all of above and costs 300/month for family. Plans seem fine.**	Window? Yes ☒ No
	Computers **in start up** — Printer **in start up**
	File Cabinets: 2 — Backup System **in start up**
Retirement program: **50 % match 403b**	Other: **office furniture provided by department. May need new chair.**
Courtesy Scholarships: **80% of tuition at RUU only**	

Moving Expenses		Information Technology	
Personal	2 months salary	Desktop Support?	yes
Lab	will move lab contents.		not an issue
		Mac/PC issue?	
Other:	must use University movers	Bioinformatics Support?	yes
		on a fee/use basis	
		Other:	

Other Issues	
School system is excellent in the immediate area.	Parking may be a problem.

Core Facilities (note accessibility, quality, and cost)			
Animal Core	good	Peptide Synthesis	ok,
Per diem rate	0.50/cage/day	Protein Expression	no
Antibody	no	Protein Sequencing	ok
Biostatistics	n/a	Proteomics	starting up
DNA Sequencing	$12/ run	Structural Biology	no
DNA Gene Array	$300 + chip	Tissue Culture	no
Mass Spectrometry	ok	Transgenic /Knockout	$5000/KO
Microscopy/Imagining	very good	Viral Vector	no
Molecular Biology	no	Other	

■ Use this worksheet to organize the decisions that you will have to make when starting your new position. The version in the CD is larger and allows you to fill in the text.

Date: 8/22/05

Choosing your start date

Got Info	Topic	Important Info		
☒	Tenure Clock	Date/Rotating **June 1**		
☒	Experiments to be completed before move: **RT-PCR assays**			
☒	Public School Start Dates **not applicable**			
☒	Other Factors **Sell condo**	done		**Closing Dec 29**
☒	**Buy house**	done		**Closing Jan 4**

Setting up the Office

Ordered	Item	Number	Cost
☒	Chairs	3	**200 each**
☒	Desk	1	**don't know**
☒	File Cabinets	2	**500 each**
☒	Small Refrigerator	1	**120**
☒	**table**	1	**400**

Computers

Ordered	Item	Number	Cost
☒	Office	**1**	**3500**
☒	Lab	**2**	**1900 each**
☒	Printer	**2**	**700 and 400**
☒	Scanner	**1**	**600**
☐	Lap Top	**1**	**3000**
☐	Back up System	**1**	**500**
☐			

Setting up the Lab

Completed	To Do List	Completed	Interviewing Staff
☐	Equipment Worksheet	☒	**Posted ad for technician**
☐	Demo Special Equipment	☐	
☐	Supplies Worksheet	☐	
☐	Negotiating with Salespeople	☐	
☐		☐	

Teaching

Got Info	Topic	Important Info
☒	Semester / Quarter System	**semester**
☐	Start dates End dates	**don't need to know yet**
☒	Graduate dates	**starts at the end of August**
☒	Classes to audit	**Medical Mycology, Dr. Kneematoad's lectures**
☐	First teaching assignment	**next year**
☐	Other:	

Service

Got Info	Topic	Important Info
☒	Start date	**Dept. Seminar program must begin organization in April of next year.**
☐	End date	
☐	Other	

Other

EQUIPMENT SET-UP WORKSHEET

▪ Use this worksheet to create a list of equipment that you wish to purchase. It will help in the overall organization of your start-up and give you an idea of what you are spending. The version in the CD is in the landscape mode and has more space for comments.

Date:

Item	Share?	Company	Catalog Num.	Features	Comments Accessories	COST	Demo?
	Yes ☐						Yes ☐
	Yes ☐						Yes ☐
	Yes ☐						Yes ☐
	Yes ☐						Yes ☐
	Yes ☐						Yes ☐
	Yes ☐						Yes ☐
	Yes ☐						Yes ☐
	Yes ☐						Yes ☐
	Yes ☐						Yes ☐
	Yes ☐						Yes ☐
	Yes ☐						Yes ☐
	Yes ☐						Yes ☐
	Yes ☐						Yes ☐
	Yes ☐						Yes ☐
	Yes ☐						Yes ☐
	Yes ☐						Yes ☐
	Yes ☐						Yes ☐
	Yes ☐						Yes ☐
	Yes ☐						Yes ☐
	Yes ☐						Yes ☐
	Yes ☐						Yes ☐
	Yes ☐						Yes ☐
	Yes ☐						Yes ☐
	Yes ☐						Yes ☐
	Yes ☐						Yes ☐
	Yes ☐						Yes ☐
	Yes ☐						Yes ☐
	Yes ☐						Yes ☐
	Yes ☐						Yes ☐
	Yes ☐						Yes ☐
	Yes ☐						Yes ☐

SUPPLY SET-UP WORKSHEET   APPENDIX 2-3

■ Use this worksheet to create a list of the supplies that you wish to purchase. It will help in the overall organization of your start-up and give you an idea of what you are spending. The version in the CD has more space.

Date: Sheet Number:

Item	Size	Company	Catalog Num.	Quantity	COST

This worksheet outline follows the format of an NIH grant application. Use the outline to organize your thoughts and provide the information that the reviewer will want to read.

Specific Aims
    Brief general statement about subject and its importance
    Relationship of general subject to what is not known
    General goals of project
    Introduction of novel reagents
    Specific questions/hypotheses to be answered/tested
    Statement of Aims

Background and Significance
    Topic 1 of need to know info to understand grant
        Subtopic breakdown
            Background info to set up hypothesis
            Hypothesis to be tested

    Topic 2 of need to know info to understand grant
        Subtopic breakdown
            Background info to set up hypothesis
            Hypothesis to be tested

Preliminary Data/Progress Report Section
    Brief summary of discoveries or subsections to come
    Title for the subsection that is informative and conclusive
        Describe the problem being solved or issue addressed
        Describe the experiment
        Address the figure or table
        Briefly describe the controls
        Draw your conclusions
        Address caveats in the interpretation and what also needs to be done or where the data are headed
        State where in the grant those experiments will be discussed or where the results were published
    Repeat Motif for next subsection

Experimental Design
    Aim 1
        Rationale for aim and design
        Approach to be taken
        Experimental detail
            Controls
        Anticipated Results
        Alternatives to get at question of interest
        Summary of Aim
    Aim 2
        Rationale for aim and design
        Approach to be taken
        Experimental detail
            Controls
        Anticipated Results
        Alternatives to get at question of interest
        Summary of Aim

This specific aims page is written about a nonsense project, but yet has all of the essential features of a real specific aims page. The example is diagrammed as described in *Chapter 3*.

A. SPECIFIC AIMS

Disease relevance and significance of problem

Individuals infused by goblegoo are profoundly affected by increases in their blahdeblah levels. In children under the age of 2, this leads to excessive Ohmygauds, culminating in decreased attention span. However, the predominance of this affliction is in the adult population, where the disease causes severe problems with unafflicted individuals who come in contact with infused individuals. Unafflicted individuals experience severe headaches upon exposure to goblegoo and excessive blahdeblah. Both individuals fail to concentrate, often resulting in accidents while driving and using the cell phone. This disease affects 20% of the adult population. Shaadupps are dominant negative regulators of blahdeblah. Shaadupps have appeared in history as far

Some history, followed by the long-term goal(s)

back as the Egyptian hieroglyphics. Despite this long history, little is known about the mechanism of Shaadupps' induction or how goblegoo is transmitted. The long-term goal of this project is to determine the molecular and cellular mechanism of Shaadupps' expression and goblegoo transmission.

Background of the system with the details that are necessary to understand the problem(s) to be solved in the application.

Cloned shortly after the invention of the family unit, Shaadupps were found to be dimerized mispronounced truncated syllables. It was recently observed that Shaadupps could be expressed in several isoforms. However, it is not clear if these isoforms are encoded by the same gene. Additionally, Shaadupps were observed to be expressed at different levels, depending on the source and strain of goblegoo, but how this level is controlled is unknown. Louder et al. suggested that Shaadupps are expressed in some individuals constitutively but that most individuals can be induced to express the gene. Moreover, the use of Shaadupps on a regular basis has been associated with a basic characteristic: impatience. Nonetheless, more and more people have begun to express Shaadupps while traveling in crowded places. It is has been proposed by Iamright et al. that the future of civilization depends on attaining an equilibrium between goblegoo, the levels of blahdeblah, and Shaadupps. In this

Here we have the gist of the proposal: the questions/hypotheses to be tested.

proposal, we seek to determine the basic parameters of this equilibrium by answering the following questions: what are the different isoforms of Shaadupps; how is their expression regulated; and do different isoforms regulate goblegoo more effectively?

> A special reagent or technique is introduced that makes the approach novel.

This investigation will take advantage of a unique set of reagents that express precise levels of blahdeblah in response to goblegoo.

> Justification for the project or one of the aims and how the work will impact the quality of life.

Because the transmission of goblegoo can occur rapidly through the population, often within minutes of its introduction, we seek to understand the mechanism of transmission. Understanding the characteristics of Shaadupps and the transmission mechanisms of goblegoo will allow the development of vaccines to prevent accidents and increase attention spans of young adults. We also hope to reduce impatience associated with excessive blahdeblah levels. Specifically, we will:

> Here the aims are listed as directed statements from which a reviewer can get a feel for the application. There are many variations on this approach. Some grants incorporate some of the background, hypotheses, and approaches into their aims. If you choose the latter route, you will have to reduce the introductory justification and rationalization remarks above. An alternative Aim 2 provides an example of this style.

Aim 1) Determine the genetic basis for the isoform variants associated with the goblegoo-induced expression of Shaadupps;

Aim 2) Determine the molecular mechanisms by which the expression of Shaadupps is controlled; and

Aim 3) Examine the strength of the different Shaadupps isoforms in curtailing blahdeblah levels, both transiently and during long-term exposure to blahdeblah.

> Alternative Style for Aim 2. Investigators use different approaches to present their aims.

Aim 2: The molecular mechanisms by which the expression of Shaadupps is controlled by blahdeblah will be determined. Due to its sensitivity to quiet-toxins, we hypothesize that Shaadupps expression is initially regulated through a Q-protein receptor and signaling cascade. Slow and rapidly transmitting mutant strains of goblegoo will be used to assess the kinetics of the Shaadupps expression and to examine the signaling path of the Q-protein receptor.

> Regardless of your style, remember that you should use one page or less to present your specific aims. You will have lots of space to justify your proposal and excite the reviewer.

NIH Detailed Budget Page

Appendix 3-3

DETAILED BUDGET FOR INITIAL BUDGET PERIOD — DIRECT COSTS ONLY					FROM	THROUGH	
PERSONNEL (Applicant Organization Only)					DOLLAR AMOUNT REQUESTED (omit cents)		
NAME	ROLE ON PROJECT	TYPE APPT. (months)	% EFFORT ON PROJ.	INST. BASE SALARY	SALARY REQUESTED	FRINGE BENEFITS	TOTALS
Ima Starr	Principal Investigator	12	50	$80,000	$40,000	$9,280	$49,280
Seymor Sells	Research Specialist III	12	100	$43,000	$43,000	$9,976	$52,976
Bean A. Student	Graduate Student	12	100	$20,000	$20,000	$0	$20,000
Wanna D. Gree	Graduate Student	12	100	$20,000	$20,000	$0	$20,000
Gott A. Kleen	Dishwasher	12	25	$18,000	$4,500	$1,044	$5,544
					$0	$0	$0
					$0	$0	$0
SUBTOTALS ⟶					$127,500	$20,300	$147,800

CONSULTANT COSTS	$0

EQUIPMENT (Itemize)		
Triplex GammaZoid	18,000	$18,000

SUPPLIES (Itemize by category)				
		Plasticware	$5,000	
PCR Supplies	$5,000	Oligonucleotides	$2,500	
Molecular Biology Supplies	$5,000	Misc. Lab Supplies	$6,700	
Radioisotopes	$2,000	Antibodies	$5,000	
Tissue Culture	$4,000	Animal Purchase	$2,500	$37,700

TRAVEL		
One meeting / year for PI and one student /fellow to present work	$2,000	$2,000

PATIENT CARE COSTS	INPATIENT	$0
	OUTPATIENT	$0

ALTERATIONS AND RENOVATIONS (Itemize by category)	
	$0

OTHER EXPENSES (Itemize by category)				
Service Contracts	$3,000			
Computer Services	1000	Flow Cytometry	$4,000	
Publication Page Charges	$1,500			
Animal Maintanence Costs	$10,000			$19,500

CONSORTIUM/CONTRACTUAL COSTS	DIRECT COSTS

SUBTOTAL DIRECT COSTS FOR INITIAL BUDGET PERIOD	$225,000

CONSORTIUM/CONTRACTUAL COSTS	FACILITIES AND ADMINISTRATIVE COSTS

TOTAL DIRECT COSTS FOR INITIAL BUDGET PERIOD	$225,000

SBIR/STTR Only: FEE REQUESTED

BUDGET FOR ENTIRE PROPOSED PROJECT PERIOD
DIRECT COSTS ONLY

BUDGET CATEGORY TOTALS		INITIAL BUDGET PERIOD (from page 4)	ADDITIONAL YEARS OF SUPPORT REQUESTED			
			2nd	3rd	4th	5th
PERSONNEL: Salary & fringe benefits Applicant organization only		$147,800	$153,712	$159,860	$166,254	$172,904
CONSULTANT COSTS		$0	$0	$0	$0	$0
EQUIPMENT		$18,000	$0	$0	$0	$0
SUPPLIES		$37,700	$39,208	$40,776	$42,407	$44,103
TRAVEL		$2,000	$2,000	$2,000	$2,000	$2,000
PATIENT CARE COSTS	INPATIENT	$0	$0	$0	$0	$0
	OUTPATIENT	$0	$0	$0	$0	$0
ALTERATIONS AND RENOVATIONS		$0	$0	$0	$0	$0
OTHER EXPENSES		$19,500	$23,500	$21,990	$20,040	$17,875
SUBTOTAL DIRECT COSTS		$225,000	$218,420	$224,626	$230,701	$236,882
CONSORTIUM/CONTRACTUAL COSTS	DIRECT	$0	$0	$0	$0	$0
	INDIRECT	$225,000	$234,000	$243,360	$253,094	$263,218
TOTAL DIRECT COSTS		$450,000	$452,420	$467,986	$483,795	$500,100

TOTAL DIRECT COSTS FOR ENTIRE PROPOSED PROJECT PERIOD (Item 8a)-> **$2,354,301**

JUSTIFICATION (Follow the budget justification instructions exactly. Use continuation pages as needed).

BUDGET JUSTIFICATION PAGE MODULAR RESEARCH GRANT APPLICATION				
Initial Budget Period	Second Year of Support	Third Year of Support	Fourth Year of Support	Fifth Year of Support
$ 225,000	$ 225,000	$ 225,000	$ 225,000	$ 225,000
Total Direct Costs Requested for Entire Project Period				$ 1,250,000

The above budget was estimated from the spread sheets in Appendix 3-2. As you can see from the 5-year plan, three of the years were below $225,000 and the last year was above. Because up to 25% of funds can be carried forward from one year to the next, this budgeting system works. You should also be aware that the funding agency is likely to cut the budget.

A copy of this form can be obtained from the NIH website.

Personnel

List the personnel who will receive a salary. State their effort on the project, what they will do, and provide some justification to have them work on the project.

Ima Starr, PhD, will serve as Principal Investigator for this project. She will devote 50% of her effort towards the completion of the aims of the project. Dr. Starr will be responsible for and will supervise the experimental plan proposed in this application.

Seymour Sells, MS, a Senior Research Specialist, grade III will devote 100% of his effort to this project. Mr. Sells joined Dr. Starr's lab two years ago and is responsible for the preliminary data for Aims 1 and 2 of the project. He has 10 years of molecular biology skills and has worked with two different animal model systems. He will focus primarily on Aim 1 of the application but will assist in all of the aims.

Bean A. Student, BS, is a second-year graduate student who has been in the lab for one year. Mr. Student developed the strains of blahdeblah described in the application and will use these in Aim 2 of the proposal. Support is required for all students after their first year in the graduate program.

Wanna D. Gree, BS, is a first-year graduate student who gained extensive laboratory experience as a research technician before entering graduate school. She will require support in her second year, which is coincident with the proposed starting date of this application. Ms. Gree will work on Aim 3. Ms. Gree contributed the western blots presented in the preliminary data section of the proposal.

Equipment

> Equipment is a one-time expense and needs to be justified. In the example, need, availability, and proximity were used as the justification. Replacement of existing instrumentation that requires constant repair can also be used. Regardless of what else you state, the success of your experiments must depend on access and use of the instrument that you request.

A **TriPlex GammaZoid** is requested to analyze the samples produced in Aims 2 and 3 of the application. The data presented in the preliminary results section were collected on Dr. Brillodooz's instrument located in the building next door to ours. However, this arrangement is not practical due to the location of the instrument and its availability. Thus, funds are requested to purchase the instrument and the two specialized accessories ($18,000). A quote is attached.

HIRING RES. TECH. WORKSHEET   APPENDIX 4-1

A full-size version of this worksheet can be found in the CD. It is arranged as a form page that can be used with each interview candidate. Use the Tab key to move between data entry zones.

Candidate's Name	Joe Petri	Date of Interview	**May 2006**
College/ Degree/ Major/year **U. South by Southeast/ B.S. Biology/ 2000**			
Years of Work Experience: **2**		Earliest/latest Start Date: **October 1**	
Area(s) of Experience:	**Molecular Biology, Animal genotyping, some subcloning**		
Future Plans (e.g. Grad School)**grad**			
Are there any reasons that you would not be able to work 40 hours a week in the lab? Yes ☐ No ☒ If so, what?			
Previous Salary: **31,000**		Range to offer: **29-36**	
Reason for seeking new position Grant Expired			

Lab Experience

In discussing his or her work experience in research laboratories, can the candidate describe details to show that he or she has actually performed certain assays, rather than just being aware that these assays exist? If so, what assays? **Can describe Southern blotting, PCR, and plasmid vector design very well.**

Can the candidate describe why certain experiments were performed in the previous job? What were they?   **Yes. He worked on the generation of mice lacking the wiggle gene. His goal was to generate homozygous animals that wobble and transfer the genotype to a different strain.**

Does the candidate's name appear on any scientific publications? Yes, 1 middle author

Comments on the skills this candidate would contribute to your lab: **The candidate would be ideal for my new transgenic mouse project and could participate in all subcloning projects.**

References

Name **Dr. Knose**	Position **Professor**	Yrs with **2** Candidate

Relationship:**Supervisor**

Rate the following from 1-5 with 1 being the best

Productivity **2**	Figure Quality Data **2**	Attendance **1**	Notebook **2**	Independence **2**
Ability to accept new direction **1**	Gets along with others **1**	Attention to Job **1**	Would you hire this person again? Yes ☒ No ☐ **Absolutely!**	

Strengths:**Good hands, hard worker**

Weaknesses: **Can leave a messy bench somedays.**

Name	**Dr. SSE Advisee**	Position	**Associate Profesor**	Yrs with Candidate	**4**

Relationship:Undergaduate advisor and honors thesis mentor

Rate the following from 1-5 with 1 being the best

Productivity **2**	Figure Quality Data **1**	Attendance **1**	Notebook **3**	Independence **2**
Ability to accept new direction **2**	Gets along with others **1**	Attention to Job **1**	Would you hire this person again? Yes ☒ No ☐ **Worked on honors thesis, would make a good graduate student**	

Strengths: **Was great in the lab. Said that he thought Joe would go to grad school. He had good hands**

Weaknesses: **Played tennis which was somewhat of a distraction from his work. However, he worked very hard in his senior year to complete his honors thesis.**

Summary

Major Plusses	Good hands, could be serious about science, and experience in transgenic animals
Major Minuses	Could be messy in the lab...
Other:	

Overall Rating: Gotta Hire ☒   **Maybe** ☐   **Don't know** ☐   **Definitely not** ☐

Salary Offer: 33,000

Other Comments:

RESEARCH UTOPIA UNIVERSITY

DEPARTMENT OF MACRO AND MICROSCIENCES

Big City, GA 30000
Tel: 404-555-9999

This letter has many of the features of an official letter that a Human Resource Department may want to have you use. Before providing any prospective employee with an offer letter, check with your departmental business manager.

May 3, 2006

Joseph Petri
27 Coliform Lane
East Bench, GA 30000

Dear Mr. Petri:

It is a pleasure for me to offer you a Research Specialist level II position in my laboratory. This position, which will begin on June 1, has a starting salary of $35,128 per annum (paid monthly) and comes with full university benefits. This includes the regular university holidays and 10 paid vacation days per year. As we discussed, because you are transferring jobs within the university, you will use up most of your accrued vacation time prior to joining the lab, and the remaining 3 days, plus the days you accrue while in my lab, will be available to you over the course of the first year of your employment. Your responsibilities will include the supervision of work-study students and the performance of experiments involving the elucidation of the molecular biology of the blahdeblah pathway. You will also be in charge of keeping radiation safety records in the laboratory.

A six-month evaluation period is required for all new hires to our department. At that time you will be eligible for a small increase in salary.

I look forward to your joining the laboratory and anticipate a productive, professional relationship.

Sincerely,

Ima Starr, PhD
Assistant Professor

The Freezer/Refrigerator Storage system is designed to help you organize and find your reagents over the course of your lifetime. Both Macintosh and PC versions written with FileMaker™ Pro version 7.0 are included. The file was designed in Dr. Boss' laboratory and is in use. Modifications have been made to make it more general. The program can be modified and expanded, which you may wish to do. Importantly, the file can be used as it is. Below is a discussion of different Views and the fields:

Form View — This is the data entry and general search view that you should use. A panel of active buttons is provided that allows you to see the two different views and box contents/grid pages.

To add a new record, press the new record button.

- **Name of Box**: Each Box should have an assigned name that you will be able to place with an individual or a reagent.
- **Box Owner**: While you own all the boxes, the person who is using the box should be the owner. In this case: Curly.
- **Location**: If you have the funds to get a storage rack system for your –80 or –20° freezer, you should do it. This allows you to have assigned space for people and to have boxes/reagents in the same place for years and years. With a freezer rack system, you can assign coordinates to the box. Here shelf 1, rack 2, row 3, column 4. Thus, s1r2r3c4.
- **Category**: Here we have added a number of categories with an edit option if you do not use any of these. Note that general lab supplies make up most of the list, with My Stuff being the only personal category.
- **Other Info**: This can be used for anything. Important info to store would be the quality of a reagent that is stored in this box or if items have been removed. For example, if the box contained competent *E. coli* cells for transformation you could write: Electrocompetent *E. coli* efficiency 1.2×10^9 colonies/µg. You could also write Stay Out or DO NOT THROW AWAY ...
- Date Created, Date Modified, and Inventory ID are automatic values.

List view — allows you to print a summary sheet for notebooks. The SORT function allows you to sort by creator, location, and content.

BOX CONTENTS view and Box Contents — provides a 10 x 10 grid for storage boxes. Using this view, –80° and liquid nitrogen freezer contents can be cataloged. Each grid is attached to a box.

Print out a copy of this page from the CD and place with your hard copy of this file.

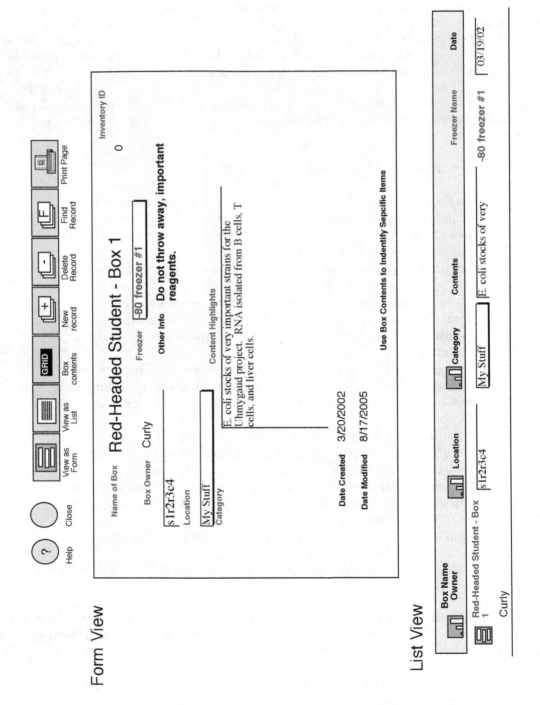

BOX CONTENTS OR GRID VIEW

	A	B	C	D	E	F	G	H	I	J
1	type in what you want								Tube 47 The good stuff	Tube 48 Much better than 47
2							my good stuff		Tube 46 not very good stuff	Nuc. Ext. Hela cells
3	E.coli XL1-Blue electro 5-6-02 1×10^9	E.coli XL1-Blue electro 5-6-02 1×10^9	E.coli XL1-Blue electro 5-6-02 1×10^9	E.coli XL1-Blue electro 5-6-02 1×10^9	E.coli XL1-Blue electro 5-6-02 1×10^9	E.coli XL1-Blue electro 5-6-02 1×10^9	E.coli XL1-Blue electro 5-6-02 1×10^9	E.coli XL1-Blue electro 5-6-02 1×10^9	E.coli XL1-Blue electro 5-6-02 1×10^9	E.coli XL1-Blue electro 5-6-02 1×10^9
4	E.coli XL1-Blue electro 5-6-02 1×10^9	E.coli XL1-Blue electro 5-6-02 1×10^9	E.coli XL1-Blue electro 5-6-02 1×10^9	E.coli XL1-Blue electro 5-6-02 1×10^9	E.coli XL1-Blue electro 5-6-02 1×10^9	E.coli XL1-Blue electro 5-6-02 1×10^9	E.coli XL1-Blue electro 5-6-02 1×10^9	E.coli XL1-Blue electro 5-6-02 1×10^9	E.coli XL1-Blue electro 5-6-02 1×10^9	E.coli XL1-Blue electro 5-6-02 1×10^9
5										
6										
7										
8										
9										
10										

The Peptide File system is designed to help you organize and find your reagents over the course of your lifetime. Both Macintosh and PC versions written with FileMaker™ Pro version 7.0 are included. The file was designed in Dr. Boss' laboratory and is in use. Modifications have been made to make it more general. The program can be modified and expanded or used as is. Below is a discussion of different views and the fields:

Peptide File View — This is the data entry and general search view that you should use. The fields are discussed.

To enter a new record, click on "New Peptide."

- **Peptide Name**: Give it a meaningful name. If you are making a series of peptides from the same protein, names like CYC27–48, meaning cytochrome c amino acids 27 through 48, are very clear. This is better than Joey23 for Joey's 23rd peptide.
- **Creator**: This is a pop-up menu that allows you to make up a list for easy entry and ID. An edit function allows you to make changes.
- **Sequence**: Enter here.
- **Gene**: Enter gene abbreviation.
- **N Term** and **C Term**: Enter the positions of the ends of the peptide. This may come in real handy when you review this material years later.
- **Region of Protein**: Indicate if the peptide is in the N-terminal region, in a DNA-binding domain, in an SH2 domain, etc...
- **Gene Reference**: Place reference information here that may be useful later.
- **Molecular Weight**: Useful for all calculations later. Often this is supplied with the peptide and is easily entered when the peptide arrives (before the label wears off or the sample is transferred).
- **Purification**: Pop-up menu to indicate if the peptide was HPLC purified or purified by some other method.
- **Concentration**: Enter the concentration that the peptide stock is now in.
- **Moles/mg**: Will be helpful later, but not necessary if you put in the molecular weight
- **Location**: Pop-up menu

List View — allows you to print all the peptides out for a complete inventory. You can also sort and find in this category as well.

- Print out this sheet from the CD to keep with your hard copies of your peptide inventory.

Peptide File

Lab Peptide File

New Peptide | Sort | Print | Find

Number 1

Date Entered 8-05-055

Creator Bossman

PeptideName **Sighting-1**

Sequence ELVISWASHERE

Gene 50s

Amino Acid Positon

N Term 1956   C Term 1967

Region of Middle
Protein

Gene Ref

Molecular Weight 1200

Purification HPLC

conc. 10   mg/ml

moles / mg

Location

Peptide Box 1

Peptide List

Lab Peptide File

New Peptide | Sort | Print | Find

PeptideName	Number	Conc.	Gene	N Term	C Term	Location
Sighting-1	1	10	50s	1956	1967	Peptide Box 1

The Plasmid File Storage system is designed to help you organize and find your reagents over the course of your lifetime. Both Macintosh and PC versions written with FileMaker™ Pro version 7.0 are included. The file was designed in Dr. Boss' laboratory and is in use. Modifications have been made to make it more general. The program can be modified and expanded or used as is. Below is a discussion of different views and the fields:

Data Entry View — This is the data entry and general search view that you should use.

- **Name of Plasmid**: Here you should assign a name that you will be able to place with an individual or a reagent.
- **Creator**: The person who made the plasmid.
- **Gene Cloned**: Obvious
- **Promoter**: If this is an expression vector, the promoter driving the gene should be inserted here.
- **Bacterial Drug Marker**: Antibiotic resistance used to select plasmid in bacteria.
- **Eukaryotic Drug Marker**: Antibiotic resistance marker for eukaryotic cells.
- **Eukaryotic Origin of replication**: If your plasmid has one, insert here.
- **Size**: Base pairs or kilobase pairs.
- **Polylinker Restriction sites**: Insert the sites here. This can be messy and you may want to have a series of templates to handle this one. This can be set up by changing the field definition to pop-up menu and defining a list.
- **Location of plasmid**: This is split into two fields, the freezer/refrigerator and which box. Box contents should be stored in a separate file.
- **Concentration of DNA**: mg/ml.
- **Location of bacterial stock**: You should keep bacterial stocks of your plasmids as it will save you time when you have to prepare fresh DNA.
- **Other name of clone**: Labs tend to give all their creations several names. List those here.
- **Date of Creation**: Enter date the clone was verified.
- **Description**: Put as much information in this category as possible as it will be important after your lab has created or collected 300 plasmids.
- **Plasmid Map**: The current picture was pasted from a DNA mapping program in Vector NTI™.

The **Sort** and **Find** buttons will help you move through the database.

List View — Allows you to print out all your plasmids with their location. You can sort these by any of the fields shown.

Print a copy from the CD for your Notebook

Data Entry

Your Laboratory Plasmid File System [New Plasmid] [Sort] [Find] [Print]

Plasmid Number 1

PlasmidName CIITA-8

Date Entered 4/20/2002

Creator J. Boss

Base Vector pREP4

Gene cloned CIITA

Date of Creation 2/21/2002

Promoter RSV

Description
Plasmid will express the CIITA gene from the CMV promoter.
Should function in most cells.

Bacterial Drug Marker amp

Eukaryotic Drug Marker hygromycin

Eukaryotic Origin EBV origin

Size 16,880

Polylinker Restriction Sites

Location of Plasmid 4° refrigerator 2   Box 1

concentration of stock (mg/ml) 2

Location of Bacterial Stock -80° freezer 2   Box 4

Other name of clone

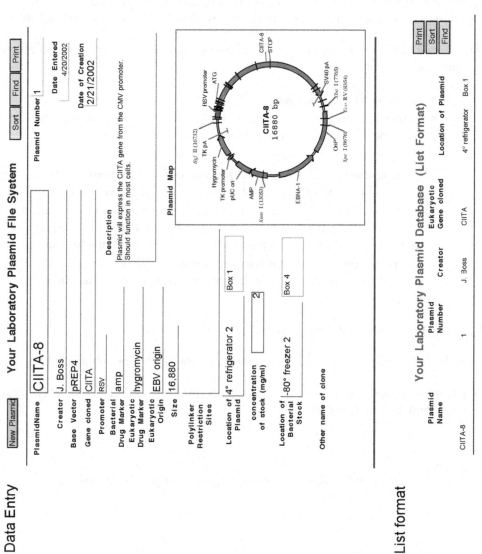

Plasmid Map

CIITA-8
16880 bp

List format

Your Laboratory Plasmid Database (List Format) [Print] [Sort] [Find]

Plasmid Name	Plasmid Number	Creator	Eukaryotic Gene cloned	Location of Plasmid
CIITA-8	1	J. Boss	CIITA	4° refrigerator Box 1

The Reagent Inventory system is designed to help you organize and find your reagents over the course of your lifetime. Both Macintosh and PC versions written with FileMaker™ Pro version 7.0 are included. The file was designed in Dr. Boss' laboratory and is in use. Modifications have been made to make it more general. The program can be modified and expanded or used as is. Below is a discussion of different views and the fields:

Data Entry View — This is the data entry and general search view that you should use. The fields are discussed below.

- **Reagent Name**: Place here
- **Category**: This is a pop-up menu that allows you to make up a list for easy entry and ID. An edit function allows you to make changes.
- **Vendor**: Also as a pop-up menu with edit features.
- **Catalog number, last size ordered, date ordered/arrived** and location serve to help you keep track of items and to allow easy reorder.
- **Inventory date** and **Current Inventory** is to allow you to view what you have at any time. This is rarely used by anyone but can be useful if you go through lots of a particular reagent that is hard to get.
- **Comment box**: Sometimes there are different types of the same reagent. Indicate here if there is any special property or use of this reagent.

To change Your Laboratory Name to your own laboratory name:

Go to layout mode, click on "Your Laboratory Name" and type what you like.

List View — allows you to print all the reagents out for a complete inventory. You can also sort and find in this category as well.

Print this sheet from the CD to place with your hard-copy records.

Data Entry View

Your Laboratory Inventory

New Record | Print | Find

Reagent Name Ethanol

Category Solvents

Card Number 0

Vendor Fisher

Catalog Number Alcohol-primo

Last size ordered 1 jug

Date Ordered 3/20/2002

Date of Arrival 3/22/2002

Box number 1

Comment box
The Good Stuff

Location in Lab Jerry's file cabinet

Date of Inventory 3/22/2002

Current Inventory half left

Academic Scientists at Work   Page Number 1   Date Printed 10/26/2005   9:30:52 AM

List View

Reagent Name	Category	Vendor	Catalog Number	Location in Lab	Current Inventory
Ethanol	Solvents	Fisher	Alcohol-primo	Jerry's file cabinet	half left

Academic Scientists at Work   Page Number 1   Date Printed 10/26/2005

Sort | Find

The Oligonucleotide Database File Storage system is designed to help you organize and find your reagents over the course of your lifetime. Both Macintosh and PC versions written with FileMaker™ Pro version 7.0 are included. The file was designed in Dr. Boss' laboratory and is in use. Modifications have been made to make it more general. The program can be modified and expanded or used as is. Below is a discussion of different views and the fields:

To enter a new oligonucleotide into the database, click on the New Oligo button.

Data Entry View — This is the data entry and general search view that you should use.

- **Name of Oligo**: Here you should assign a name that you will be able to place with a gene.
- **Designer**: The person who designed and ordered the oligo.
- **Gene**: Obvious.
- **Purpose**: Describe the use of the oligo. Primer, probe, etc.
- **Comments**: Describe any special comments, such as the appropriate primer pair for PCR or if the oligo is modified, etc.
- **Sequence**: Enter as read from the tube/order, not what was intended. This is a good time to check to make sure that what was ordered is correct.
- **Concentration, T_m, and nmol/A260**: These values will help you in your calculations and if you do them once, you won't have to repeat them over and over again.
- **Location of oligo**: This is a pop-up menu to allow you to have storage boxes.

The **Sort** and **Find** buttons will help you move through the database.

List View — Allows you to print out all your oligos with their location. You can sort or use the find buttons to limit the set.

> Print a copy of this page from the CD to keep with the hard copy of your Oligonucleotide database.

Entry Layout

New Oligo   *Your Laboratory Oligonucleotides*   Sort | Find | Print

Blah-5-245   OligoName   **Designer** Curly

OligoNumber 1   **Date** 8/02/02   Comments

Manufacturer Life Sciences   **Gene** Blah-2   Use with primer Blah 3-245 to amplify entire coding region.

Purpose PCR primer for coding region of Blahdeblah -2 gene

Sequence
ATG CCC GGC TCC GTG GGG CAA   Concentration 1.0
TAT TTT CAC   mg/ml

                                                                 Tm

                                                      nmol/260

Location Oligo box 2

List view

Sort | Find | Print

OligoName	OligoNumber	Designer	Sequence	Concentration	Gene
Blah-5-245	1	Curly	ATG CCC GGC TCC GTG GGG CAA TAT TTT CAC	1.0	Blah-2

The Antisera Database File Storage system is designed to help you organize and find your reagents over the course of your lifetime. Both Macintosh and PC versions written with FileMaker™ Pro version 7.0 are included. The file was designed in Dr. Boss' laboratory and is in use. Modifications have been made to make it more general. The program can be modified and expanded or used as is. Below is a discussion of different views and the fields:

To enter a new antiserum into the database, click on the "New Record" button.

Data Entry Views 1 and 2 — There are two data entry views provided. All record fields are shared such that information typed into one record is in the other.

- **Name of Antisera**: Here you should assign a name that you will be able to place with a protein.
- **Creator of page**: The person who entered the data.
- **Company, Catalog number,** and **Lot Number**: These parameters are important if you will be reordering the same antisera. Lot number is VERY important in case a different lot does not provide you with the same data.
- **Date Tested, Location of Western (or other assay)**, and the recommended **standard dilution** from that assay should be listed here. This will save you lots of time if this data is efficiently recorded.

Because some antisera are homemade, other important pieces of information should be documented. Including

- **Immunogen description** and **location**: Enter what was used to immunize the animal.

Often homemade antibodies are purified to enrich the signal or reduce background. Some common methods are listed.

- ***E. coli* cleared** indicates that the antiserum was passed over a column bound with bacterial proteins.
- **Ig purified** indicates that the antibodies were purified with a protein A or protein G resin.
- **Affinity purified** indicates that the antibodies were purified over a peptide or protein column. **Date and Notebook page** for these experiments should be listed so that you can go back a year or two later and refresh your memory about the quality of the antisera.

The **Sort** and **Find** buttons will help you move through the database.

List View — Allows you to print out all your oligos with their location. You can sort or use the find buttons to limit the set.

> ▪ Print a copy of this sheet from the CD for your notebook.

Data View 1

Lab Antisera Catalog System

| New Record | Print Page | | Sort | Find |

Name of Antisera **Shaadupp-N**

Creator of page Student with the Curly red hair

Date Entered 3-25-2002

Company Rockland Farms   Cat. Number   Lot Number 21

Type of Antisera Rabbit

Location Big Freezer   Box 11

Location of Data from company Antisera notebook

Date Tested 4-1-02

Location of Western Curly notebook 7 pg 65

Standard Dilution? 1:1000

If We Made it section

Immunogen Description PEPTIDE - SHAADVPPPELVISWASHERE

Location of Immunogen Big Freezer   Box 12

	Date	Notebook page
E.coli Cleared ☒	4-10-2002	Curly 7 pg 65
Ig purified ☒	4-10-2002	Curly 7 pg 66
Affinity purified ☐		

List view

Lab Antisera Catalog System

| Print Page |

Name of Antisera	Type of Antisera	Location	Lot Number	E.	Ig	A	Company	Cat. Num.
Shaadupp-N	Rabbit	Box 11	21	☒	☒	☐	Rockland	

Location

Name of Antisera Shaadupp-N

Lot Number 2 1

Type of Sera Rabbit

Bleed Type: ⊙ test ⊙ production ⊙ terminal

Box 11

Company Rockland Farms

Cat. Number

Purification -

	Date	**Location in Notebook**
E.coli Cleared	4-10-2002	Curly 7 pg 65
Ig purified	4-10-2002	Curly 7 pg 66
Affinity purified		

Location of Western Curly notebook 7 pg 65   **Date Tested** 4-1-02

Standard Dilution? 1:1000

Immunogen Description PEPTIDE - SHAADVPPPELVISWASHERE

Location of Immunogen Box 12

Location of Data from comany Antisera notebook

Creator of page Student with the Curly red hair

Use this when people are leaving your lab group to ensure that their work is properly documented and that the data that were generated in your lab can be used by others and for future publications. Because it can take weeks for someone to document their reagents and work, be sure to go through the list with your departing personnel at least 2 weeks before they leave.

DATA

Notebooks		
		All notebooks should have their pages numbered
		All notebooks should have a <u>table of contents</u> (TOC) in the front that lists the experiment title, date and page.
		A master notebook should be formed that includes all of the TOCs from each notebook. It will also include the items listed below.
E-data		Redundant data sets should be removed.
		Data pertaining to each manuscript should be in its own folder.
		Manuscripts in preparation should be in separate files, with ALL their associated data.
		All data and files should be recorded onto a CD and a duplicate made.

REAGENTS

Plasmids		
		All Plasmids should be clean and free of DNase. (CsCl purified is best, phenol extracted mini-preps will suffice.)
		All plasmids stocks should be stored in 2 ml ring sealed tubes.
		The tubes should be clearly labeled both around the sides and on the top.
		Plasmids need to be entered into the lab plasmid database. Please include as much detail as possible so that the construction design will be known.
		E. coli stocks of plasmids should be organized and stored at –80°C.
		A table of contents of each plasmid box (4°, -20°, and –80°) should be prepared and placed with the master TOC.
Cells		All Cell lines should be frozen and stored in liquid nitrogen.
		Vials should be clearly labeled with the date and passage number
		A table or grid for each liquid nitrogen box should be placed with the master TOC.

Antibodies		If you ordered antibodies, they should be placed in the common antibody racks in the lab. The specification sheet that came with the antibody should be in your master notebook.
Oligonucleotides		All stocks should be placed in the general laboratory racks and assigned a number.
		Sequences, name, and number should be entered into the lab oligo file.
Peptides		All stocks should be placed in the general laboratory racks
		Sequences, name, and number should be entered into the lab peptide file
Mouse strains		A complete description of your transgenic/ knockout mouse strains, crosses, and constructs should be included in the master notebook.
		PCR Assays for following genetic markers should be listed in a separate section

Other		

Please also include this page in the master notebook along with your forwarding address, phone, and e-mail.

GOALS AND PROGRESS WORKSHEET

A blank copy of this worksheet can be found in the CD.

Date: August 20, 2002

Funding

Goal	Write grant on Blahdeblah regulation in the immune system.
Progress	Grant submitted to NIH and American Cancer Society
Goal	Convert Aim 1 of NIH grant to institutional seed grant
Progress	Funded for 1 year at $25,000

Lab Staff

Goal	Hire a Research Specialist and attract a student
Progress	Hired Suzy Hood, and

- The student with the curly red hair is rotating through the lab

Lab Set-Up

Goal	Purchase GammaZoid and small equipment
Progress	Tested Triplex and DuPlex instruments. Decided to go with the Triplex due to footprint size, service, and the fact that Dr. N.X. Door has one too.

Research

Goal	Establish cell lines expressing blahdeblah.

- Clone UM-7 into expression vector.
- Test role of the UM-7 in culture system.

Progress	Experiments are up and running. Waiting on the GammaZoid to collect data. Feasibility of the project looks good and will be preparing supplement to my NIH grant before study section meets.

Papers

Goal	Complete manuscript with postdoctoral advisor and submit
Progress	Submitted to RealScience
Goal	Revise grad student paper
Progress	Experiments completed, get it out of here!

Teaching

Goal	Attend lectures in Medical Microbiology
Progress	Attending Dr. Brillodooz's lectures on tolerance

Service

Goal	Organize Department Seminar Program
Progress	Organizing the seminar program with the help of Dr. Didit. We have Dr. Hotshot coming next semester.

> ▪ This manuscript outline contains many aspects of a scientific paper. Because journals vary on the order or titles for the different sections, be sure to check with the "Notes to the Authors" section for specific variations in the format. This outline can also be used with your students to explain how to write a paper. One exercise for your students is to have them diagram a manuscript using this outline.

I. **Title Page**
    a.   Title
    b.   Title Page information
    c.   Running Title

II. **Abstract/Summary**
    a.   General statement and setup of system
    b.   Question or hypothesis to be tested
    c.   Results
    d.   Conclusions

III. **Introduction**
    a.   General Statements about the system
    b.   Importance of the system
    c.   What is known
    d.   What is controversial or undefined
    e.   Hypothesis or questions that are to be addressed in the manuscript
    f.   Approach
        i.   Special reagents
        ii.   Technology
    g.   High points of the paper and overall conclusions

IV. **Materials and Methods**
    a.   Reagents
        i.   Cells
        ii.   Antibodies
        iii.   Nucleic acids
        iv.   Etc
    b.   Procedures
        i.   Brief description or reference to procedure
        ii.   Details of new procedure or new way of using a procedure

V. **Results**
    a.   Overall setup or defining the system
    b.   First series of experiments
        i.   Why is the experiment being performed?
            1.   Testing an hypothesis
            2.   Answering a question

    ii. The approach
      1.   Special reagents
      2.   Technology or procedure used
    iii. The result — the Figure or Table
    iv. Interpretation of the result
      1.   this suggests (demonstrates or proves) that
      2.   alternative interpretations
  c. Second series of experiments
    i. Why?
      1.   Distinguishing between the alternative explanations?
      2.   The above data suggest ...
    ii. Repeat the motif in B.
    iii. More figures and tables
  d. Additional results sections

VI. **Discussion**
  (Below are potential items to consider including in your manuscript. Note not all will be applicable.)
  a. Summary of the high-points (this is usually presented, although some find it repetitive; thus make it short)
  b. Model developed from experimental data (if applicable)
    i. May include a figure
  c. Explanation of Model
  d. Alternative interpretations of model
  e. Justification for your interpretation
  f. Interesting conclusions about your work
    i. Relationship of your conclusions to the work of others
  g. Change in paradigm due to your results
  h. Other interesting conclusions about your work
  i. Placement of your work in the grander scheme of biology (this can also go at the beginning)

VII. **Acknowledgments**
  a. Those who provided some guidance to the work
  b. Those who provided reagents
  c. Agencies that provided support (very important)

VIII. **Literature Cited**
  a. (be sure to check journal format)

IX. **Figure Legends**
  a. Figure 1
    i. Concluding statement about the figure
    ii. Description of the experiment and any information that may not have been in the materials and methods
    iii. Conclusions based on the data. This is requested by some journals
  b. Figure 2, etc.

X. **Tables**
  a. Tables do not have figure legends so they must be clearly marked
  b. Use superscripts to define columns, rows, or items that will be unclear

XI.   **Figures**

a.   Should be large enough so that the reviewer and the reader can see the data

b.   Lettering should be at least one size larger than you think it should be so that when it is reduced, it will still be easy to read. Use san serif fonts.

c.   Label the figures on the back so that the reviewer and publisher can tell which is which.

d.   If you have photos or autorads, print these on high quality paper or provide real photographs for the reviewers. After all, you want them to be able to appreciate the beauty of your data.

RESEARCH UTOPIA UNIVERSITY
DEPARTMENT OF MACRO AND MICROSCIENCES
Big City, GA 30000
Tel: 404-555-9999

August 12, 2005

Dr. Reed N. Wright
Editor
Top of the Heap Journal
1600 King Hill
Washington, DC 00090

Dear Dr. Wright:

> In the first section, describe the results and their implications. Try to point out what is important and how the work affects the field.

Please accept for review and publication in *Top-of-the-Heap Journal* our manuscript entitled: "Blahdeblah over expression leads to a loss of motor control." Using a novel series of transgenic animals that over express wild-type and mutant blahdeblah proteins, simplified motor control was tested. The results of the experiments showed that animals expressing high levels of blahdeblah had almost no motor control. When these animals were crossed to the Shaadupp dominant negative animals, normal responses were observed in the offspring. The use of the transgenic animals and these results provide a direct link between high levels of blahdeblah and the loss of motor control. The results alter the current paradigm on the mechanism of blahdeblah function and have implications for treating humans with excess blahdeblah levels.

> You have the opportunity to suggest reviewers. You should choose individuals who will be able to critically review your manuscript.

The following scientists could effectively review this manuscript: Dr. J.D. Talkker, University of the North by Northwest; Dr. E.Z. Reviewer, State University; and Dr. R.N. Ace, University of the South by Southeast. Due to competition on this project, I request that Dr. Gonna Scooptcha at Big Time University not be asked to review the manuscript.

Thank you for considering our manuscript for your journal and I look forward to a positive and rapid review.

Sincerely,
Ima Starr, PhD

Assistant Professor

■ Consider using a form like this for your students to provide you with feedback for your course. You should note that the evaluation attempts to be an objective review of the course and stays away from personality issues. If there are multiple instructors in the course, use a separate form for each one. A blank copy of this form is in the CD.

Course Name: Real-Time Science   Semester: Fall 2005
Instructor: Dr. Binbusi Elsewhere

DIRECTIONS: Please mark your opinion of this course on the scale provided	Strongly Agree 1	2	3	4	Strongly Disagree 5	N/A
1. Course content (assignment, lectures/seminars, etc.) was congruent with overall course objectives.				xxx		
2. Expectations were clear regarding assignments.				xxx		
3. Time spent on assignments was reasonable for the number of credit hours.					xxx	
4. Course materials (AV media, handouts, etc.) were helpful for achieving course objectives.						xxx
5. Teaching methods were appropriate for the material presented.					xxx	
6 Teaching methods were helpful to you, considering your learning style.					xxx	
7. Class fostered interaction and sharing of knowledge and ideas.				xxx		
8. Class objectives clearly defined.					xxx	
9. Feedback on assignments was timely.	xxx					
10. Overall, this was an excellent class.					xxx	

What were the strengths of the course? Dr. Elsewhere found the room on time. He seems to be smart and is possibly a leader in his field. The material is very important. Nice tie.

Suggestions for improvement: Find out what was covered in the course.

Other Comments: Dr. Elsewhere appeared to have hastily prepared his lectures and did not seem to know what we had done the day before as most of his lectures repeated Dr. Starr's outstanding presentation.

> • To improve upon your own teaching style and to remind yourself of what you thought of your course, take a few minutes after you have completed your teaching assignment and fill out this form. Be sure to describe any problems that you had or any ideas you have on how to improve the course/lectures for the next time that you have to teach. A blank copy is available on the CD.

Course Title: Concepts in Immunology
Term Dates: 2005
Role in Course: Director

DIRECTIONS: Rate the quality of your development and teaching of this course on the scale provided.	Strongly Agree				Strongly Disagree	N/A
1. The objectives of the course are clear.	☒	☐	☐	☐	☐	☐
2. The course assignments are relevant to the lectures and exams.	☐	☒	☐	☐	☐	☐
3. The textbook/reading assignments are valuable for the students.	☐	☒	☐	☐	☐	☐
4. I explain complex topics clearly.	☒	☐	☐	☐	☐	☐
5. Exam questions are clear and students are appropriately evaluated.	☒	☐	☐	☐	☐	☐
6. I am well prepared for lectures and questions from the class.	☒	☐	☐	☐	☐	☐
7. I am effective in generating student participation in class.	☒	☐	☐	☐	☐	☐
8. Feedback on assignments was timely.	☐	☐	☒	☐	☐	☐
9. Students meet with me to discuss the course.	True ☒			False ☐		
10. Percentage of students who drop the course	15 %					

11. What are the most valuable aspects of this course? **The course provided a broad overview of immunology. Most topics were covered in sufficient detail that the students would be able to follow most seminars.**
12. What would you change next time (or for the rest of the term)? I **would spend more time on the molecular aspects of gene regulation.**

13. What improvements would you make in the way you teach and organize this course?
Course organization/Syllabus: **Find a replacement for Dr. Kneematoad. Add additional section on dendritic cells. Expand innate immunity section.**
Textbook/Assigned reading: **Be sure to point out to the students where the material is**

	presented in the book.
	Office hours for meetings with students: **Have more defined office hours. Perhaps Friday afternoons.**
	Examinations: **Go to more of a short answer format instead of all multiple guess. Mix up the format a little. Because class size is small the TA can help grade the essays.**
	Other student assignments: **Incorporate a paper into the course.**

This document should serve as a detailed record of your teaching experiences and contributions. Because you will forget your contributions over time, it is important to begin this document early and update it once a year. A blank version is located in the CD.

Student Training

This is not just a log of whom you have trained, but it also tracks where they have gone. The better they do in their careers, the more valuable your training record.

Graduate Students				
Student	**Degree (Yr)**	**Thesis**	**Postdoc Mentor/Institution**	**Current Position**
Bill Emsa	PhD (00)	Alleles of Ohmygaud receptors	Kent Believeit, Big H Univ.	Assist. Prof. U. Obewon
James Jones	PhD (01)	Cloning and characterization of Ohmygaud	Reese P. Torr	Post Doc Cytokine Design Labs
Jean Mann	current	Binding of Ohmygaud to Shaadupps		

Post Doctoral Trainees			
Fellow	**Years in Training**	**Project**	**Current Position**
Frank D. P'doc	1995-1999	Designing chemical inhibitors of Ohmygaud	Asst. Prof. Skywalker University
Bea Stinyears	2001-present	Regulation of Ohmygaud binding proteins	

Undergraduate Trainees			
Student	**Degree (yr)**	**Honors Thesis**	**Current Position**
Katie Bee	BA (2000)	Ohmygaud mutants in yeast	Medical Resident

Student Thesis Committees

Student thesis committees (other than those of your own students) require a substantial time commitment and serve to educate the students about their dissertation work.

Student	**Term**	**Degree (yr)**	**Program**
Phil O. Sophie	94-97	PhD (97)	Genetics
Frank Aurnot	96-98	MD/PhD (99)	MSTP
Bill Meelater	00- present		Immunology
Johnny B. Good	02- present		Immunology
Benny Serve	03- present		Genetics

Student Qualifying Exams

> Graduate programs approach qualifying exams in numerous ways. Some programs have a standing committee while others use the thesis committee. Each uses your time and counts as teaching. List your efforts in graduate programs here.

2002 — Genetics Qualifying Exam committee. Participated in three oral and written examinations.

2005 — Invited external examiner for the oral qualifying exam of Robert Watchout at the Utopia University, Graduate Program in Immunology.

2000, 2001, 2002, 2003, 2004, 2005 — Immunology Qualifying Exam committee. Member of Oral exam committee. Examined 5–8 students.

CLASSROOM TEACHING

> List the years that you taught, the name of the course, with course number, who the course was taught to, and your exact role.

Professional Courses

1998, 1999, 2000 — Medical Genetics, PAG501. This course is taught to 1st-year physician assistants. Taught 2 lectures on the genetic basis of immunodeficiencies.

2001, 2002, 2003, 2004, 2005 — Medical Microbiology and Immunology, MMI555. This course is taught to the 2nd year medical students. Taught 4 lectures on antibodies, T cell receptors each year.

Graduate Courses

2000, 2004 — Eukaryotic gene expression, EGE720. This current topics course was taught to 2nd, 3rd and 4th year graduate students. Topics included RNA polymerase and general transcription factors, transcription elongation, gene specific transcription factors, and the role of chromatin in gene expression. Taught and organized entire course.

Undergraduate Courses

2001 — Basic Immunology, BI422. This course is taught yearly to senior undergraduates and is a survey course of the field of immunology. Taught 12 one-hour lectures covering the topics of antigens, antibodies, helper and cytotoxic T cells, peripheral tolerance, and autoimmunity.

BEDSIDE TEACHING

> Physician scientists can spend a substantial portion of their time training medical students and residents at the bedside. Typically they will be involved with a certain aspect of their training. List the general section of the student's training and the number of contact hours.

2004 — Hematology and Oncology rotation: approximately 30 contact hours with students and patients.

NATIONAL TEACHING

> As described in the text, there are many ways in which you can gain a national teaching reputation. Below are some examples. If you have accomplished any of these, you should include them here.

TEXTBOOKS

Approaches to cloning genes. Editor: J. Allright. Lockstone Press. 2005.
      Chapter 1. "Introduction to cloning animals" Ima Starr
      Chapter 3. "Expression vector design and use" Ima Starr

EXAM COMMITTEES

2003–present — Member National Board Examination Committee for Transplantation

WORKSHOPS

2003 — Lecturer, Using adenovirus vectors for expression cloning. Molecular Biology Workshop at Summer Research Conference, Lockstone Resorts. 3 1-hr lectures.

2004 — Lecturer, Design of viral vectors for gene expression and therapy. VLS (Very Large Society) Annual Meeting, Orlando, FL.

PUBLIC EDUCATION

Public Education Workshops
2003 — Lecturer, Use of Viral Vectors in Gene Therapy. Seminar to local investor groups.

Videos

2002 — *The Attack of the Transgenes*. Consulted with WKBBT studios on the development of an educational video.

Media

2001 — Interviewed for public television by WKBBT, Channel 86, on the topic of "Cloning of animals for spicy food."

Ima Starr
Assistant Professor
Today's date: August 8, 2005

National Service: (■current)

Grant Application Review

> List in chronological order the past and current review groups in
> which you participated. Indicate the term that you served. State
> whether you are/were a full standing member or an ad hoc member
> (i.e., served only once).

National Institutes of Health; Ad Hoc Member: Cellular and Molecular Immunology A
Study Section, Jan '05.

American Cancer Society; Member: Study Section for Cancer and Immunology August '01
to July '04

■National Institutes of Health, NIGMS, Predoctoral Training Grants, October 2004 to '07.

Journal Reviewer

> List the journals for which you serve(d) on the editorial board
> first. Under the reviewer category list all journals for which you
> served as a reviewer.

Editorships
Associate Editor, *The Journal of Immunology*, April 2005–present.

Reviewer
*Science, Nature, Cell, EMBO, Immunity, Immunogenetics, Journal of Biological Chemistry,
Journal of Immunology, Molecular and Cellular Biology.*

Scientific Meeting Organizer
Co-Organizer; Southeastern Conference on Ohmygaud Response, August 2002.

University Service: (■ current)

> List the duties that you have been assigned. Use chronological
> order and includes the tie period that you served. Services that
> required extensive time commitment should be listed under
> separate categories.

Core Laboratory Service:
DNA Synthesis Facility: Faculty supervisory Committee. 2003–2004

Graduate Program Service
▪Member, Executive Committee for Genetics and Molecular Biology Graduate Program. Spring 2003 to present.

▪Member, Oral Qualifying Exam Committee for Genetics and Molecular Biology Graduate Program. 2004 to present.

University Committees
Chair, Department of Microbiology Space Committee: 2005.

Member, Faculty Search Committee, Department of Pathology, 2004.

Member, Departmental Program Retreat Committee, 2004.

Clinical Service

> Describe your clinical service. This can be a description of your practice or your laboratory oversight.

Director, HLA-Typing Facility — 2004–present. Responsibilities include overseeing the typing of tissues for four area hospitals. Patient load of the facility is approximately 2000 samples per year.

Pediatrician, St. Gracious Hospital — 2000-present. Service includes two afternoons a week. Patient load averages, 1000 patient visits/yr.

REFERENCES AND RESOURCES

HOW TO WRITE SCIENTIFIC PUBLICATIONS

Alley, M. (1997). *The Craft of Scientific Writing*, Third Edition. New York: Springer-Verlag Inc.

Booth, V. (1993). *Communicating in Science: Writing a Scientific Paper and Speaking at Scientific Meetings*, 2nd Edition. Cambridge, UK: Cambridge University Press.

Day, R.A. (1998). *How to Write and Publish a Scientific Paper*, 5th Edition. New York: Greenwood Publishing Group Inc.

Day, R.A. (1995). *Scientific English: A Guide for Scientists and Other Professionals*, 2nd Edition. Washington, DC: Oryx Press.

Gopen, G.D., & Swan, J.A. (1990, Nov–Dec). The Science of Scientific Writing. *American Scientist*, **78**, 550–558.

Huth, E.J. (1999). *Writing and Publishing in Medicine*. New York: Williams and Wilkins.

Rankin, R. (2001). *The Work of Writing: Insights and Strategies for Academics and Professionals*. San Francisco: Jossey-Bass.

Yang, J.T., & Yang, J.T. (1995). *An Outline of Scientific Writing for Researchers with English as a Foreign Language*. New Jersey: World Scientific Publishing Co. Inc.

HOW TO GET A JOB AND WHAT TO DO WHEN YOU GET IT

Association of American Medical Colleges. (Published Annually). *Report on Medical School Faculty Salaries*. Washington, DC: Association of American Medical Colleges.

Barker, Kathy. (2002). *At the Helm, A Laboratory Navigator*. New York: Cold Spring Harbor Laboratory Press.

Bianco-Mathis, V., & Chalofsky, N. (1998). *The Full-time Faculty Handbook*. Thousand Oaks, CA: Sage Publications.

Boice, R. (2000). *Advice for New Faculty Members*. Needham Heights, MA: Allyn and Bacon Inc.

Fiske, P. (1996). *To Boldly Go ... A Practical Career Guide for Scientists*. Washington, DC: American Global Publishing.

Heiberger, M.M., & Vick, J.M. (2001). *The Academic Job Search Handbook*, 3rd Edition. Philadelphia: University of Pennsylvania Press.

Lanks, K.W. (1990). *Academic Environment: A Handbook for Evaluating Faculty Employment Opportunities*. New York: Hemisphere Publishing Corp.

Medawar, P.B. (1990). *Advice to a Young Scientist*. New York: Basic Books.

Menges, R. (1999). *Faculty in New Jobs: A Guide to Settling In, Becoming Established and Building Support*. San Francisco: Jossey-Bass Inc.

Reis, R.M. (1997). *Tomorrow's Professor: Preparing for Academic Careers in Science and Engineering*. New York: Institute of Electrical and Electronics Engineers Inc.

HOW TO PREPARE FOR TENURE REVIEW

American Association of University Professors (AAUP). (2001). *Policy Documents and Reports*. 9th Edition. Washington, DC: AAUP.

Commission on Academic Tenure in Higher Education. (1973). *Faculty Tenure: A Report and Recommendations*. San Francisco: Jossey-Bass.

Diamond, R.M. (2004). *Preparing for Promotion and Tenure Review: A Faculty Guide*, 2nd Edition. Boston, MA: Anker Publishing Co.

Toth, E. (1997). *Ms. Mentor's Impeccable Advice for Women in Academia*. Philadelphia: University of Pennsylvania Press.

How to Manage Your Time

Harvard Business School Essentials. (2005). *Time Management: Increase Your Personal Productivity and Effectiveness*. Boston: Harvard Business School Press.

Lakein, A. (1996). *How to Get Control of Your Time and Your Life*. New York: Signet.

Mackenzie, A., & Mackenzie, R.A. (1997). *The Time Trap: The Classic Book on Time Management*. New York: AMACOM.

Mancini, M. (2003). *Time Management*. New York: The McGraw-Hill Companies.

Other Useful References and Resources

American Council on Education, American Association of University Professors, and United Educators. (2000). *Good Practice in Tenure Evaluation: Advice for Tenured Faculty, Department Chairs, and Academic Administrators*. Washington, DC: American Council on Education.

Baldwin, R.G. (1990). Faculty career stages and implications for professional development. In J.H. Schuster, D.W. Wheeler & Associates (Eds.), *Enhancing Faculty Careers: Strategies for Development and Renewal*. San Francisco: Jossey-Bass.

Bauer, D.G. (1999). *The "How To" Grants Manual: Successful Grant-seeking Techniques for Obtaining Public and Private Grants*, Fourth Edition. Phoenix: The American Council on Education.

McGuire-Dunn, C., & Chadwick, C. (2004). *Protecting Study Volunteers in Research: A Manual for Investigative Sites*, 3rd Edition. Rochester, NY: CenterWatch.

McKeachie, W.J., & Hofer, B.K. (2002). *McKeachie's Teaching Tips: Strategies, Research, and Theory for College and University Teachers*, 11th Edition. Boston: Houghton Mifflin.

Truss, L. (2003). *Eats, Shoots & Leaves: The Zero Tolerance Approach to Punctuation*. New York: Penguin Group.

Software Citations

FileMaker Pro (2004). Santa Clara, CA: Claris Corporation.

Microsoft Word (2003). Redmond, Washington: Microsoft Corporation.

Microsoft Excel (2003). Redmond, Washington: Microsoft Corporation.

Statistical Packages for the Social Sciences (SPSS), Version 10.07. (2000). Chicago: SPSS Inc.